NATIONAL STANDARD OF THE PEOPLE'S REPUBLIC OF CHINA

Technical Standard for Ice and Snow Landscape Buildings

GB 51202 - 2016

Chief Development Department: Ministry of Housing and Urban-Rural Development of the People's Republic of China

Approval Department: Ministry of Housing and Urban-Rural Development of the People's Republic of China

Implementation Date: July 1, 2017

China Architecture & Building Press

Beijing 2016

图书在版编目(CIP)数据

冰雪景观建筑技术标准(英文)/住房和城乡建设部组织编译. —北京：中国建筑工业出版社，2019.7
ISBN 978-7-112-23640-4

Ⅰ.①冰… Ⅱ.①住… Ⅲ.①景观-建筑工程-技术标准-英文 Ⅳ.①TU986-65

中国版本图书馆CIP数据核字(2019)第075350号

Chinese edition first published in the People's Republic of China in 2017
English edition first published in the People's Republic of China in 2019
by China Architecture & Building Press
No. 9 Sanlihe Road
Beijing, 100037
www. cabp. com. cn

Layout Designed in China by Beijing Yan Lin Ji Zhao Co. , LTD
© 2016 by Ministry of Housing and Urban-Rural Development of
the People's Republic of China

All rights reserved. No part of this publication may be reproduced or transmitted in any form or by any means, graphic, electronic, or mechanical, including photocopying, recording, or any information storage and retrieval systems, without written permission of the publisher.

This book is sold subject to the condition that it shall not, by way of trade or otherwise, be lent, re-sold, hired out or otherwise circulated without the publisher's prior consent in any form of blinding or cover other than that in which this is published and without a similar condition including this condition being imposed on the subsequent purchaser.

ISBN 978-7-112-23640-4(33926)

Announcement of Ministry of Housing and Urban-Rural Development of the People's Republic of China

No. 1333

Announcement on Issuing the National Standard *Technical Standard for Ice and Snow Landscape Buildings*

Technical Standard for Ice and Snow Landscape Buildings has been approved as a national standard with a serial number of GB 51202−2016. It shall be implemented on July 1, 2017, of which Articles 4. 3. 2, 4. 3. 5, 4. 3. 8, 5. 1. 3, 5. 5. 5 are compulsory provisions and must be enforced strictly. *Technical Specification for Ice and Snow Landscape Buildings* (JGJ − 247 − 2011) shall be abolished simultaneously.

Authorized by Standard Quota Research Institute of our ministry, this Standard is published and distributed by China Architecture & Building Press.

Ministry of Housing and Urban-Rural Development of the People's Republic of China
October 25, 2016

Foreword

According to the requirements of Document Jian Biao [2013] No. 169 issued by Ministry of Construction (MoC)——"Notice on Printing the Development and Revision Plan of National Engineering Construction Standards in 2014". The drafting committee, after conducting extensive investigation and research, summarizing practical experience carefully, referring to international standard and foreign advanced standard, has revised the standard on the basis of soliciting opinions extensively.

The main technical contents of the standard include: 1. General Provisions; 2. Terms and Symbols; 3. Ice and Snow Material Calculation Indicators; 4. Design of Ice and Snow Landscape Buildings; 5. Construction of Ice and Snow Landscape Buildings; 6. Construction of Power Distribution and Illumination; 7. Acceptance Check; 8. Maintenance Management.

The provisions in bold type in the standard are compulsory and must be implemented strictly.

The Ministry of Housing and Urban-Rural Development of the People's Republic of China is responsible for the management and interpretation of the compulsory provisions of the standard; Harbin Architectural Designing Institute is responsible for the explanation of the specific technical provisions. For any comments and suggestions in the course of implementation, please contact Harbin Architectural Designing Institute (Address: No. 117 Youyi Road, Daoli District, Harbin, Postcode: 150010).

Chief Editorial Units:
- Harbin Architectural Designing Institute
- HarbinWujian Construction Engineering Co., Ltd.

Joint Editorial Units:
- Harbin Academy of Supervision Association
- Harbin Urban and Rural Construction Committee
- Harbin Municipal Survey and Design Association
- Harbin Institute of Technology
- Harbin Modern Group Co., Ltd.
- China Construction Engineering Design Group Corporation
- Harbin Sai Ge Yin Xiang Design Company
- Harbin Cultural Tourism Group Co., Ltd.
- Academy of Urban-Rural Planning and Design, Sun Yan-sen University
- Harbin Zhong Tai Enterprise Management Co., Ltd.
- Heilongjiang College of Construction

Main Drafters:

Hao Gang	Shen Baoyin	Tang Rongbin	Chen Jiliang
Liu Baizhe	Wang Lisheng	Wang Dongtao	Cheng Yan
Che Xingbin	Sun Jingzhong	Wei Fengping	Wang Wenyu
Yang Hongwei	Cao Shengxuan	Peng Junqing	Ma Xinwei

Tao Chunhui	Zhao Zeyuan	Wu Yudan	Xia Qianming
Liu Ruiqiang	Sun Ying	Liu Yang	Yang Fushuang
Zhang Lining	Zhang Shoujian	Su Yikun	Gao Yang
Qi Yilin	Hao Jia	Shen Kai	Jiang Zhiping
Zhao Weixia	Lv Mifeng	Wang Tongjun	Wu Gang
Pu Wenzheng	Ma Zhe	Wu Zhiguo	Dong Chenming
Wu Xiaosong	Su Enming	Chen Xudong	Wang Zijun
Yue Qiang	Tian Lichen	Sun Guimin	

Main Examiners:

Wang Gongshan	Zhu Weizhong	Yang Shichang
He Zhendong	Shi Jiaxiang	Li Tao
Xiang Wei	Lv Bin	Li Xinze
Ma Yan	Guo Keguang	

English Translators:

Ma Xinwei	Wang Junping	Hao Jia	Pu Wenzheng
Hao Gang	Liu Xiaoping		

English Reviser:

Malcolm A. Perry (U.K)

Contents

1 General Provisions ………………………………………………………………………… (1)
2 Terms and Symbols ……………………………………………………………………… (2)
 2.1 Terms ………………………………………………………………………………… (2)
 2.2 Symbols ……………………………………………………………………………… (3)
3 Ice and Snow Material Calculation Indicators ………………………………………… (4)
 3.1 Ice Materials ………………………………………………………………………… (4)
 3.2 Snow Materials ……………………………………………………………………… (5)
4 Design of Ice and Snow Landscape Buildings ………………………………………… (8)
 4.1 General Requirements ……………………………………………………………… (8)
 4.2 Scenic Area Planning and Design ………………………………………………… (8)
 4.3 Architectural Design ………………………………………………………………… (9)
 4.4 Structural Design of Ice Masonry ………………………………………………… (11)
 4.5 Structural Design of Snow Construction ………………………………………… (17)
 4.6 Illumination Design of Snow and Ice Landscapes ……………………………… (22)
 4.7 Intelligentization Design …………………………………………………………… (28)
5 Construction of Ice and Snow Landscape Buildings ………………………………… (29)
 5.1 General Requirements ……………………………………………………………… (29)
 5.2 Construction Survey ………………………………………………………………… (29)
 5.3 Ice-collecting and Snow-making …………………………………………………… (30)
 5.4 Foundation Construction of Ice Building ………………………………………… (30)
 5.5 Construction of Ice Masonry ……………………………………………………… (31)
 5.6 Construction of Steel Structure in Ice Masonry ………………………………… (32)
 5.7 Construction of Watered Icescape ………………………………………………… (32)
 5.8 Ice Sculpture Making ……………………………………………………………… (33)
 5.9 Ice Lantern Making ………………………………………………………………… (33)
 5.10 Snowscape Building Construction ………………………………………………… (34)
 5.11 Snow Sculpture Making …………………………………………………………… (34)
6 Construction of Power Distribution and Illumination ………………………………… (36)
 6.1 Construction of Power Distribution Cable ……………………………………… (36)
 6.2 Illumination Construction …………………………………………………………… (37)
7 Acceptance Check ………………………………………………………………………… (40)
 7.1 General Requirements ……………………………………………………………… (40)
 7.2 Acceptance Check of Dominant Items of Ice Masonry ………………………… (41)
 7.3 Acceptance Check of General Items of Ice Masonry …………………………… (42)
 7.4 Acceptance Check of Dominant Items of Snow Masonry ……………………… (43)
 7.5 Acceptance Check of General Items of Snow Masonry ………………………… (44)
 7.6 Acceptance Check of Power Distribution and Illumination …………………… (44)

8　Maintenance Management	(46)
8.1　Monitoring	(46)
8.2　Maintaining	(46)
8.3　Dismantling	(47)
Appendix A　Influence Coefficients of Bearing Capacity of Ice Masonry	(48)
Appendix B　Influence Coefficients of Bearing Capacity of Snow Masonry	(49)
Appendix C　Records of Engineering Quality Acceptance	(50)
Appendix D　Division Works of Ice and Snow Landscape Buildings	(57)
Explanation of Wording in This Standard	(58)
List of The Quoted Standards	(59)
Addition: Explanations of Provisions	(61)

1 General Provisions

1.0.1 The standard B formulated for the purpose of improving the level of design, construction, inspection, maintenance and management of ice and snow landscape buildings and guaranteeing advanced technology, safety, reliability, energy conservation, environmental protection to ensure the acceptable engineering quality.

1.0.2 The standard B applicable to the design, construction, acceptance, maintenance and management of ice and snow landscape buildings with ice and snow as the main building materials.

1.0.3 The design, construction, acceptance, maintenance and management of ice and snow landscape shall comply with the standard and also the provisions of current national standard.

2 Terms and Symbols

2.1 Terms

2.1.1 Ice and snow landscape buildings

The recreational facilities and artistic landscapes with ice and snow artistic features such as ice lanterns, snow sculptures, ice sculptures and ice and snow architectures.

2.1.2 Natural ice

The ice that is naturally formed in rivers and lakes in low temperature environment.

2.1.3 Man-made ice

The ice that is formed under the artificial refrigeration conditions.

2.1.4 Rough ice

Natural ice that has not been cut purposefully.

2.1.5 Ice collecting

The process of using machines to cut apart the natural ice with certain standard, and obtain rough ice.

2.1.6 Ice masonry

Elements like walls and columns used for ice landscape buildings that are made by laying ice blocks by standard and freezing them with water.

2.1.7 Watered icescape

Ice scenery that is made bymechanically or manually spraying water on the skeletons with certain shapes and freezing them.

2.1.8 Ice flowers

The icescape in which plants, flowers, fruits, fishes or insects are frozen by putting them in a mold that is full of water.

2.1.9 Ice sculpture

A sculpture that is made of ice.

2.1.10 Ice lanterns

The icescape with artistic effects made by carving hollow ice masonry, which is created by injecting water into molds or containers using artificial refrigeration and subsequently putting lights into the lantern.

2.1.11 Natural snow

Natural snowor perennial snows in nature.

2.1.12 Man-made snow

Tiny ice crystals that are made of water using special equipment in a low temperature, or tiny ice particles made of ice shattered with special equipment.

2.1.13 Snow sculpture

A sculpture that is made of snow.

2.1.14 Rough snow masonry

Geometric shapes of compacted snow with certainstandard and strength.

2.1.15 Height of ice or snow sculpture buildings

The height from the outdoor ground to the top of ice masonries or snow masonries of ice and snow landscape buildings.

2.2 Symbols

2.2.1 Material performance

f an ice or snow masonry's design value of compressive strength;

f_t an ice or snow masonry's design value of axial tensile strength;

f_{tm} an ice masonry's design value of bending tensile strength;

f_v an ice masonry or snow masonry's design value of shearing strength;

f_w a snow masonry's design value of bending tensile strength.

2.2.2 Actions and its effects

M design value of cross section bending moment;

N design value of the axial pressure;

N_L design value of axial force in a local compression area;

N_t design value of axial tension;

V the design value of cross section shearing force.

2.2.3 Geometric parameters

A cross-sectional area of a member;

A_L local compression area;

H height of a member;

H_0 calculated height of a wall or column;

h thickness of walls or short side length of rectangular columns;

S spacing between transverse walls;

W sectional resistance moment of a member.

2.2.4 Calculating coefficient

φ the effect-coefficient of bearing capacity;

β ratio of height to sectional thickness of walls or columns;

$[\beta]$ permissible ratio of height to sectional thickness of walls or columns.

3 Ice and Snow Material Calculation Indicators

3.1 Ice Materials

3.1.1 The ultimate values of compressive strength, tensile strength and shearing strength of ice shall follow those shown in Table 3.1.1.

Table 3.1.1 The ultimate values of compressive strength, tensile strength and shearing strength of ice (MPa)

Strength Type	Ice Block Temperature Grading (℃)					
	−5	−10	−15	−20	−25	−30
Compressive strength	2.790	3.090	3.510	4.050	4.710	5.490
Tensile strength	0.108	0.109	0.111	0.114	0.119	0.125
Shearing strength	0.360	0.450	0.550	0.640	0.740	0.830

3.1.2 The standard values of compressive strength, tensile strength and shearing strength of ice masonries shall follow those shown in Table 3.1.2.

Table 3.1.2 The standard values of compressive strength, tensile strength and shearing strength of ice masonries (MPa)

Strength Type	Ice Masonry Temperature Grading (℃)					
	−5	−10	−15	−20	−25	−30
Compressive strength	0.854	0.946	1.075	1.240	1.442	1.681
Tensile strength	0.047	0.047	0.047	0.048	0.049	0.050
Shearing strength	0.078	0.088	0.097	0.105	0.112	0.119

3.1.3 The design values of compressive strength, axial tensile strength and shearing strength of ice masonries shall follow those shown in Table 3.1.3.

Table 3.1.3 The design values of compressive strength, axial tensile strength and shearing strength of ice masonries (MPa)

Strength Type	Characteristics of damage	Ice Masonry Temperature Grading (℃)					
		−5	−10	−15	−20	−25	−30
Compressive strength	Regular masonry section	0.475	0.526	0.597	0.689	0.801	0.934
Axial tensile strength	Along the ice masonry and its saw-tooth joint section	0.026	0.026	0.026	0.027	0.027	0.028
Shearing strength	Along the continuous seam and the saw-tooth joint section	0.043	0.049	0.054	0.058	0.062	0.066

Note: 1　The regular masonries in the table refer to those made by bonding small blocks together with water;

　　　2　The bonding strength of the water between ice blocks is the same as the design value of ice masonries at the same temperature;

　　　3　The strength design value of the branches of battened hollow ice walls shall be taken based on 90% of the value in Table 3.1.3.

3.1.4 The ice friction coefficient, linear expansion coefficient, average density and coefficient of

heat transmission of ice shall meet the following requirements:
1 The ice friction coefficient (μ) shall be 0.1;
2 The coefficient of linear expansion (α) shall be 52.7×10^{-6}/K;
3 The ice average density (ρ) shall be 920kg/m³;
4 The ice heat transmission coefficient (λ) shall be 2.30W/(m·K).

3.2 Snow Materials

3.2.1 The density values of snow masonries shall follow the values in Table 3.2.1.

Table 3.2.1 Density values of snow masonries

Snow Type	Loose Condition	Molding Pressure		
		0.05MPa	0.10MPa	0.15MPa
Man-made snow	455kg/m³	510kg/m³	530kg/m³	550kg/m³
Natural snow	190kg/m³	350kg/m³	390kg/m³	410kg/m³

Note: The density values of the snow masonries molded under other pressures shall be obtained with an interpolation method based on the values in the table.

3.2.2 The ultimate value, normal value and design value of compressive strength of snow masonries shall follow those shown in Table 3.2.2.

Table 3.2.2 The ultimate value, normal value and design value of compressive strength of snow masonries (MPa)

Snow Type	Density (kg/m³)	Category of Values Taken for Compressive Strength	Temperature Grading				
			−10℃	−15℃	−20℃	−25℃	−30℃
Man-made snow	510	Ultimate value	0.369	0.405	0.441	0.487	0.534
		Normal value	0.199	0.218	0.238	0.263	0.288
		Design value	0.105	0.115	0.125	0.138	0.151
	530	Ultimate value	0.535	0.578	0.621	0.729	0.838
		Normal value	0.289	0.312	0.335	0.393	0.452
		Design value	0.152	0.164	0.176	0.207	0.238
	550	Ultimate value	0.701	0.751	0.801	0.971	1.142
		Normal value	0.378	0.405	0.432	0.524	0.616
		Design value	0.199	0.213	0.227	0.276	0.324
Natural snow	350	Ultimate value	0.189	0.236	0.284	0.304	0.324
		Normal value	0.102	0.128	0.153	0.164	0.175
		Design value	0.054	0.067	0.081	0.086	0.092
	390	Ultimate value	0.349	0.402	0.456	0.548	0.640
		Normal value	0.188	0.217	0.246	0.295	0.345
		Design value	0.099	0.114	0.129	0.156	0.182
	410	Ultimate value	0.429	0.485	0.542	0.670	0.798
		Normal value	0.231	0.262	0.292	0.361	0.430
		Design value	0.122	0.138	0.154	0.190	0.226

3.2.3 The ultimate value, normal value and design value of flexural strength of snow masonries shall follow those in Table 3.2.3.

Table 3.2.3 The ultimate value, normal value and design value of
flexural strength of snow masonries (MPa)

Snow Type	Density (kg/m³)	Category of Values Taken for Flexural Strength	Temperature Grading				
			−10℃	−15℃	−20℃	−25℃	−30℃
Man-made snow	510	Ultimate value	0.150	0.248	0.346	0.386	0.426
		Normal value	0.076	0.125	0.175	0.196	0.216
		Design value	0.040	0.066	0.092	0.103	0.114
	530	Ultimate value	0.288	0.436	0.584	0.632	0.680
		Normal value	0.146	0.221	0.296	0.320	0.345
		Design value	0.077	0.116	0.156	0.169	0.181
	550	Ultimate value	0.426	0.624	0.822	0.878	0.934
		Normal value	0.216	0.316	0.416	0.445	0.473
		Design value	0.113	0.166	0.219	0.234	0.249
Natural snow	350	Ultimate value	0.147	0.152	0.157	0.160	0.162
		Normal value	0.074	0.077	0.080	0.081	0.082
		Design value	0.039	0.041	0.042	0.043	0.043
	390	Ultimate value	0.223	0.235	0.246	0.255	0.263
		Normal value	0.113	0.119	0.125	0.129	0.133
		Design value	0.059	0.063	0.066	0.068	0.070
	410	Ultimate value	0.389	0.404	0.418	0.422	0.425
		Normal value	0.197	0.204	0.212	0.213	0.215
		Design value	0.104	0.108	0.111	0.112	0.113

3.2.4 The ultimate value, normal value and design value of splitting tensile strength of snow masonries shall follow those in Table 3.2.4.

Table 3.2.4 The ultimate value, normal value and design value of
splitting tensile strength of snow masonries (MPa)

Snow Type	Density (kg/m³)	Category of Values Taken for Splitting Tensile Strength	Temperature Grading				
			−10℃	−15℃	−20℃	−25℃	−30℃
Man-made snow	510	Ultimate value	0.093	0.106	0.113	0.120	0.121
		Normal value	0.047	0.054	0.057	0.061	0.061
		Design value	0.025	0.028	0.030	0.032	0.032
	530	Ultimate value	0.146	0.160	0.170	0.182	0.185
		Normal value	0.074	0.081	0.086	0.092	0.094
		Design value	0.039	0.043	0.045	0.049	0.049
	550	Ultimate value	0.194	0.205	0.216	0.228	0.231
		Normal value	0.098	0.104	0.109	0.115	0.117
		Design value	0.052	0.055	0.058	0.061	0.062
Natural snow	350	Ultimate value	0.066	0.071	0.076	0.079	0.081
		Normal value	0.033	0.036	0.038	0.040	0.041
		Design value	0.017	0.019	0.020	0.021	0.022
	390	Ultimate value	0.102	0.108	0.111	0.115	0.118
		Normal value	0.052	0.054	0.056	0.058	0.060
		Design value	0.027	0.029	0.030	0.031	0.031
	410	Ultimate value	0.149	0.162	0.170	0.177	0.183
		Normal value	0.075	0.082	0.086	0.090	0.093
		Design value	0.040	0.043	0.045	0.047	0.049

3.2.5 The ultimate value, normal value and design value of shearing strength of snow masonries shall follow those in Table 3.2.5.

Table 3.2.5 The ultimate value, normal value and design value of shearing strength of snow masonries (MPa)

Snow types	Density (kg/m³)	Category of Values Taken for Shearing Strength	Temperature Grading				
			−10℃	−15℃	−20℃	−25℃	−30℃
Man-made snow	510	Ultimate value	0.268	0.336	0.404	0.472	0.540
		Normal value	0.131	0.165	0.198	0.231	0.265
		Design value	0.066	0.083	0.099	0.116	0.133
	530	Ultimate value	0.362	0.439	0.515	0.587	0.659
		Normal value	0.177	0.215	0.525	0.288	0.323
		Design value	0.089	0.108	0.126	0.144	0.162
	550	Ultimate value	0.515	0.573	0.630	0.688	0.745
		Normal value	0.252	0.281	0.309	0.337	0.365
		Design value	0.162	0.141	0.155	0.169	0.183
Natural snow	350	Ultimate value	0.068	0.070	0.072	0.081	0.089
		Normal value	0.033	0.034	0.035	0.040	0.045
		Design value	0.017	0.017	0.018	0.020	0.023
	390	Ultimate value	0.145	0.164	0.183	0.190	0.196
		Normal value	0.073	0.082	0.090	0.093	0.096
		Design value	0.037	0.041	0.045	0.047	0.048
	410	Ultimate value	0.179	0.190	0.200	0.211	0.221
		Normal value	0.088	0.093	0.098	0.103	0.108
		Design value	0.044	0.047	0.049	0.052	0.054

4 Design of Ice and Snow Landscape Buildings

4.1 General Requirements

4.1.1 The design of ice and snow landscapes shall follow the principles of safety, environmental friendliness, economy and artistry.

4.1.2 The design of ice and snow landscape buildings shall include:

1 Overall design and special design of supporting facilities such as road, electric power, water supply and drainage, communication, etc;

2 The design of the ice and snow landscapes of architecture and art categories;

3 The structural design of ice and snow masonries;

4 The design of ice and snow landscape illuminating;

5 The design of ice and snow activities;

6 The design of supporting facilities for the service of scenic spot such as management, business, sanitation, medical and ambulance;

7 Individual landscape lighting, power distribution, sound design.

4.1.3 The dyes for making colored ice and colored snow shall be environmentally friendly and pollution-free.

4.1.4 Water supply shall meet the needs of the water consumption of ice making, snow making, construction, living, fire-fighting, etc.

4.1.5 The design of the ice and snow landscape buildings shall meet the needs of material utilization, equipment maintenance, construction operations and tourists' activities under cold conditions.

4.2 Scenic Area Planning and Design

4.2.1 The site selection shall follow the following provision:

1 The scenic area shall be reasonably planned and scientifically and conveniently sited, with fresh air, and no dust or smoke pollution, away from residential areas, with comprehensive consideration given to such factors as climate, geology, physiognomy, electrical power, telecommunications, traffic, ice sources, snow making, water sources, etc;

2 The scenic area shall be designed to meet the exhibitive function requirements, with large parking lots available, and be easy for people to gather and evacuate;

3 The site shall facilitate construction and meet the construction safety and environmental protection requirements.

4.2.2 The overall planning for the scenic area shall define the planning programs for major items or overall design thoughts.

4.2.3 The overall planning for the scenic area shall define functional division, transportation system, tour routes, auxiliary projects and corresponding signs. The area occupied by the scenic area shall be determined based on not less than 10m² per tourist at peak hours. The ice and snow consumption, power consumption and investment shall be estimated. The overall planning design achievements shall include the scenic area location map, the present situation diagram; the overall planning diagram; the overall effect drawing; the functional division diagram; the external transport organization planning diagram; scenic area's internal road transport planning diagram, personnel evacuation organization, ice collection location and transport route, location of water source for ice making and snow making, overall lighting and lighting color analysis chart and technical and economic indicators.

4.2.4 The detailed design planning for the scenic area construction shall - according to the overall plan - determine the theme, contents of each functional division, and propose the originality, position, amount of space occupied, function and technical design requirements of each ice and snow landscape. The detailed planning and design result shall include the subdivision planning diagram, the detailed scenic area construction drawings, division effect drawing, vertical design drawing, visual analysis drawing of the scenic area, tour route map, scenic area recreational activity diagram, service facilities and signage system diagram, illumination distribution diagram, background music distribution map, electric power distribution diagram and planning instructions.

4.2.5 Considering the traffic flow and people flow in peak tour hours and static and dynamic traffic organization, the traffic planning shall propose a plan for guiding people flow direction and evacuation, a plan for vehicle parking by category and canalized traffic organization and emergency plan for people and vehicle evacuation in case of sudden incidents, and determine road width, parking lot area and traffic signs.

4.2.6 The scenic area shall be equipped with emergency signs, with emergency evacuation escape and fire escape planned, emergency plan and disaster relief measures taken.

4.2.7 The scenic area shall plan and reserve charging facilities required by electric vehicles.

4.3 Architectural Design

4.3.1 Design of ice and snow landscapes of architecture category shall meet the following requirements:

　　1　The design shall guarantee the safety and functions of all structures;

　　2　The scheme design shall include horizontal drawing, vertical drawing, sectional drawing, effect drawing, rough ice and snow masonry drawing, lighting effects and technical and economical indicators; a three-dimensional model, if necessary, shall be established for important landscape buildings;

　　3　The construction drawing design shall include a site plan, building construction drawing, structure construction drawing, illumination power distribution drawing, and other special item, special design and description, material and equipment statistic list and related safety and technical measures;

　　4　The inside of the masonry of a massive ice landscape building can be designed to be hollow or filled with rough ice or crushed ice, and then watered layer by layer. The laying thickness of outside ice wall blocks shall be determined by calculation and annotated in the construction drawing.

4.3.2 Structural design shall be done for an ice and snow landscape building which has a height of more than 10m or allows tourists to go inside for activities or has load exposed on top.

4.3.3 The ice staircase shall go through anti-skid treatment, with the width of the steps not less than 350mm, and height not more than 150mm; the stairs shall lean inward, and the relative height difference shall not exceed 10mm. The height of ice handrails of the ice staircases and platforms shall not be less than 1200mm, and the thickness shall not be less than 250mm, with anti-slip warning signs provided.

4.3.4 The height of ice masonry buildings shall not exceed 30m. The height of snow masonry buildings shall not exceed 20m. The ice masonry building whose length exceeds 30m shall preferably have expansion joints whose width is not less than 20mm.

4.3.5 The masonry structures which tourists can directly touch shall be tucked or made into a staircase when their vertical height exceeds 5m. Meanwhile, the following requirements shall be observed:

    1 **Measures shall be taken against overturn and slippage;**

    2 **The thickness of ice masonries shall not be less than 800mm and be constructed in layers. The cohesion-ratio of the seam shall be no less than 80%;**

    3 **The thickness of snow masonries shall not be less than 900mm and shall be tamped in layers according to the design density value;**

    4 **The distance between the vertical projection of the highest masonry or overhanging part and the outer edge of the landscape's foundation shall be no less than 600mm.**

4.3.6 The design of ice and snow landscapes of art category shall comply with the following requirements:

    1 Distinct themes and obvious intention and clear framework;

    2 Demonstration in an exaggerated way with a prominent overall image. The relationship between the local part and the snow building as a whole must be distinctive;

    3 The amount of space occupied shall be proper. Art effects shall be prominent with good viewing effects under various illuminations. The ice and snow artistic landscapes shall be easy to carve;

    4 The gravity of works shall overlap with the form center. In case of gravity center displacement, stability technical measures that coordinate with the artistic effect of the works shall be taken.

4.3.7 The main surface of ice and snow sculptures and colored ice walls shall be opposite, fully or partially, to natural light (sunlight). It shall not directly face the prevailing wind direction. The front elevation and the back elevation of the snow landscape shall avoid direct sunlight when it's higher than 15m. If necessary, sheltering measures shall be taken. For a large snow landscape building, environmentally friendly jelly sun screen shall be sprayed over the surface facing the sunlight.

4.3.8 The design of Ice and snow recreational activity item category shall follow the following requirements:

    1 **If the height of ice and snow buildings for climbing surpasses 5m, safety devices shall be incorporated, and security-tested devices shall be available. The maintenance devices for safety, an evacuation platform and channel shall be required on the top of the building.**

    2 **The slideway of ice and snow slides shall be plain and smooth. In addition, the following requirements shall be met:**

        1) **The width of the straight slideway shall not be less than 500mm, and the width of curve**

slideway shall not be less than 600mm; the slideway guardrail shall not be lower than 500mm, and the thickness shall not be less than 250 mm;

    2) The slideway guardrail around the bend shall be heightened and reinforced. The guardrail in the curve parts shall not be less than 800mm. Caution signs shall be fixed in the turning slope change area. Buffer shall be provided at the slope end, and the length of the buffer shall be determined by field testing. At the ice-slide's end, safety facilities shall be available;

    3) For sliding activities, skating apparatus shall be applied when the slide length is longer than 30m. The average slope of the slideway without skating apparatus shall not exceed 25° and that with skating apparatus shall not exceed 10°;

    4) The skating apparatus shall be made of durable light material and prove qualified through safety testing.

3 For special recreational activities involving ice bicycles, snowmobiles, ice bumper cars and snow bumper cars, safe and eligible products shall be used and protective facilities shall be provided.

4.3.9 The design scenic area service supporting facilities shall meet the following requirements:

1 Barrier-free facilities shall be available at the entrance and exit, major roads, and service facilities of the scenic area. For the platforms, roads, stairs and rampways, where the traffic is heavy and people are easy to push, squeeze or slip, shall be equipped with anti-skid and protective facilities such as carpets, handrails, etc.;

2 For the service rooms for commerce, catering, toilets, rest, recreation, etc., equipment rooms such as the power distribution room and snow machine room and management rooms such as custom service center, ticketing center and management center, they shall be rationally laid out based on function and landscape requirements. The housing facilities shall have thermal insulation function, and their shape and material shall coordinate properly with the surroundings; for the ice and snow activity items involving skating apparatus, it shall be preferred to provide tourist and tool traction devices;

3 The service radius of commercial rooms can be 100m~150m. The service radius of public toilets can be 50m~100m.

4.4 Structural Design of Ice Masonry

4.4.1 The structure elements of ice masonries shall be designed based on the ultimate state of the bearing capacity, and meet the requirements under normal conditions.

4.4.2 When the structure elements of ice masonries are designed based on the ultimate state of the bearing capacity, calculation shall be conducted based on the most unfavorable combination of the following formulas:

$$\gamma_0(1.2S_{Gk} + 1.4\gamma_L S_{Q1k} + \gamma_L \sum_{i=2}^{n} \gamma_{Qi}\psi_{ci}S_{Qik}) \leqslant R_d \qquad (4.4.2\text{-}1)$$

$$\gamma_0(1.35S_{Gk} + 1.4\gamma_L \sum_{i=1}^{n} \psi_{ci}S_{Qik}) \leqslant R_d \qquad (4.4.2\text{-}2)$$

Where: γ_0——structure importance coefficient, 1.0 taken;

        γ_L——for variable loads, considering service life adjustment coefficient in structure design, only limited to floors and roofing, 0.9 taken;

S_{GK} —— effect of permanent load standard value;

S_{Q1k} —— effect of a variable load standard value that plays a controlling role in basic combinations;

S_{Qik} —— effect of the ith variable load standard value;

R_d —— design value of resisting force of structure elements;

γ_{Qi} —— subentry coefficient of the ith variable load, 1.4 taken;

ψ_{ci} —— coefficient of the combination value of the ith variable load, 0.7 taken.

4.4.3 The calculation of structure elements of ice masonries shall comply with the following provisions:

1 The bearing capacity of structure elements of ice masonries shall be calculated based on the ice masonry strength value at $-5°C$ in the temperature grading;

2 The dead weight of ice masonries shall be $9.2kN/m^3$;

3 The values of dead weight of non-ice masonry structure elements and the acting load shall be taken according to the relevant provisions of the current national standard *Load Code for the Design of Building Structure* GB 50009.

4.4.4 The force modes in the design of ice and snow landscape buildings shall be dominated by compression, with other force modes like tensile and shear stress to be reduced.

4.4.5 The foundation design of ice landscape buildings shall meet the following requirements:

1 For ice buildings higher than 10m, and the shorter landing edge exceeding 6m, the foundation design shall be required. The bearing capacity of the foundation shall be calculated based on non-frozen soil strength, with consideration given to the frozen expansion coefficient of the soil around it, with corresponding anti-frozen expansion measures taken;

2 For the foundation that can't meet the requirements of design, measures shall be taken to conduct foundation reinforcement, increase the bearing capacity of the foundation and at the same time, measures to improve the overall rigidity of ice masonries shall be taken;

3 As for frozen soil foundation, underlying layer calculation and foundation frozen expansion stability calculation shall be conducted.

4.4.6 For ice landscape building which has a height of less than 10m frozen soil ground foundation can be adopted, which can be accomplished by freezing the watered soil. When the thickness of frozen soil is larger than 400mm, 400mm shall be taken as the design value; when less than 400mm, the actual value of the thickness shall be adopted. The bearing capacity of frozen soil shall be determined through the in-situ test. The underlying layer calculation and foundation frozen expansion stability calculation of frozen soil foundation shall be conducted.

4.4.7 For the ice masonries, static force calculation shall be conducted in accordance with the current national standard *Code for Design of Masonry Structures*. GB 50003 Generally, the ice masonry shall be considered as a rigid structure.

4.4.8 When the ice masonry structure is regarded as a rigid structure and the overall stability (overturn-resistance, slipping-resistance, etc.) needs to be checked based on the most unfavorable combination of the following formulas:

$$\gamma_0 (1.2S_{G2k} + 1.4\gamma_L S_{Q1k} + \gamma_L \sum_{i=2}^{n} S_{Qik}) \leqslant 0.8S_{G1k} \qquad (4.4.8\text{-}1)$$

$$\gamma_0 (1.35S_{G2k} + 1.4\gamma_L \sum_{i=1}^{n} \psi_{ci} S_{Qik}) \leqslant 0.8S_{G1k} \qquad (4.4.8\text{-}2)$$

Where: S_{G1k} —— positive effect of permanent load standard value;

S_{G2k} —— negative effect of permanent load standard value.

4.4.9 The bearing capacity of a structural member in compression, shall meet the following formula:

$$N \leqslant \varphi f A \qquad (4.4.9)$$

Where: N —— design value of axial pressure;

φ —— The influence coefficient of β (the ratio of height to thickness) and e (eccentricity of axial force) to elements in compression shall follow the provisions in Appendix A of the standard. The value of β shall be taken according to Articles 4.4.14-1 and 4.4.12-2 of the standard; when e is calculated based on the endogen force design value, it shall be no less than 60% of the distance from cross section gravity center to the edge of axial eccentric section;

f —— the design value of compressive strength of ice masonries, to be taken according to the provisions in Table 3.1.3 of the standard;

A —— area of section. Ice masonry shall be calculated based on the net cross section; the flange widths of the wall sections with pilasters and ice structure columns shall follow the Items 1 and 2 of Artide 4.4.14-2 of the standard respectively; net section lengh shall be taken for the walls between pilasters or ice structure columns.

4.4.10 The bearing capacity of local compression shall suit the following formula:

$$N_l \leqslant 1.2 f A_l \qquad (4.4.10)$$

Where: N_l —— design value of axial force in a local compression area;

f —— an ice masonry's design value of compressive strength, taken based on the provisions in Table 3.1.3 of the standard;

A_l —— local compression area.

4.4.11 The bearing capacity of a structure element subject to axial tension shall suit the following formula:

$$N_t \leqslant f_t A \qquad (4.4.11)$$

Where: N_t —— design value of axial tension;

f_t —— design value of tensile strength of an ice masonry, taken based on the provisions in Table 3.1.3 of the standard;

A —— section area, calculated based on the net cross section of an ice masonry.

4.4.12 The bearing capacity of an element subject to shearing force shall suit the following formula:

$$V \leqslant f_v A \qquad (4.4.12)$$

Where: V —— the design value of shearing force;

f_v —— design value of the shearing strength of an ice masonry, taken based on the provisions in Table 3.1.3 of the standard;

A —— section area, calculated based on the net cross section area for an ice masonry.

4.4.13 The bearing capacity of member subject to bending force shall suit the following formula:

$$M \leqslant 0.8 f_{tm} W \qquad (4.4.13)$$

Where: M —— design value of cross section bending moment;

f_{tm} —— design value of the bending strength of and ice masonry, design value of bending strength can be taken, taken based on the provisions in Table 3.1.3 of the standard;

W——ice masonry's resistance moment.

4.4.14 The height to thickness ratio of walls and columns shall follow the formulas below:

1 The height to thickness ratio of ice walls and columns shall suit the following formula:

$$\beta = \frac{H_0}{h} \leqslant [\beta] \tag{4.4.14-1}$$

Where: H_0——calculated height of walls and columns, with reference to Table 4.4.14-1;

h——thickness of wall or the shorter edge of the rectangle column;

$[\beta]$——the acceptable ratio of height to thickness of walls and columns, with reference to Table 4.4.14-2.

Table 4.4.14-1 Calculated height of walls and columns H_0

Ice Mansory Building and Element Category	Floor or Roof Category	Spacing between Transverse Walls S (m)	Wall with Pilaster or Wall with Ice Structure Columns or Wall Connected with Others		
			$S>2H$	$2H \geqslant S > H$	$S \leqslant H$
Rigid scheme for ice masonry buildings	Fabricated light-duty floor and roof with purlin	$S<20$	$1.0H$	$0.4S+0.2H$	$0.6S$
	Wooden or light steel roof of tile roofing	$S<16$			
Non-rigid scheme for ice masonry buildings	Fabricated light-duty floor and roof with purlin	$S \geqslant 20$	$1.5H$		
	Wooden or light steel roof of tile roofing	$S \geqslant 16$			
Upper end is free			$2.0H$		

Note: 1 In the base course, H is the distance from top floor (or the level of supporting points) to bottom bearing of the element; on other floors, H is the distance between floors or other horizontal supporting points;

2 When the upper end of element is free, H is the length of the element;

3 For the gable without pilaster, H is to be the height between floors plus a half of the gable; for the gable with pilaster or ice column, H is the height of the gable wall at the place of the pilaster or ice column;

4 For the three-side supporting walls without cover, H is the distance from the upper free edge to bottom bearing point, and there shall be an ice ring beam, pilaster or ice construction column.

Table 4.4.14-2 Permissible ratio of height to thickness of walls and columns $[\beta]$

Member	Ice Wall	Ice Column
Main load-carrying member	10	8
Minor load-carrying member	12	10

2 The ratio of height to thickness of walls with pilaster and ice structure columns shall be verified and calculated based on the following formula:

$$\beta = \frac{H_0}{h'} \leqslant [\beta] \tag{4.4.14-2}$$

Where: H_0——calculated height of walls with pilaster, walls with ice structure columns or walls between pilasters, walls between ice structure columns shall follow the provisions in Table 4. 4. 14-1 and Table 4. 4. 14-2, Subitem 3;

h'——the converting thickness of walls with pilaster, walls with ice structure columns shall be in accordance with Subitems 1 and 2 of Article 4. 4. 14-2, and the thickness of walls between pilasters and walls between ice structure columns shall be that of the walls;

$[\beta]$——the permissible ratio of height to thickness of walls and columns shall follow the Table 4. 4. 14-2.

1) The converting thickness of the wall with pilaster shall be 3. 5 times the gyration radius of the section. For the strip foundation of the walls with pilaster, the flange width of the section of the wall with pilasters can be the distance between adjacent pilasters. For single- story ice landscape buildings, flange width of the section of the wall with pilasters can be the width of pilaster plus 2/3 of wall height, but it shall not exceed the width of the wall between windows or the distance between two adjacent pilasters. For multi-story ice landscape buildings, when there is opening between windows, flange width of the section of the wall with pilasters can be the width of ice wall; for walls without door and window openings, the width of the flange wall on each side can be 1/3 of pilaster height and shall not exceed the distance between two adjacent pilasters.

2) The flange width of the section of the wall with ice structure columns can be the distance between two adjacent ice structure columns. The converting thickness shall be 1. 05 times the thickness of the wall.

3) During the verification and calculation of the ratio of height to thickness of the wall between pilasters or between ice structure columns, spacing S between transverse walls shall be the distance between pilasters or structure columns; calculated height H_0 of the wall with pilasters or ice structure columns that has an ice ring beam shall follow Table 4. 4. 14-1, but element height H shall be defined according to the following rules: when the ice ring beam b is not less than 1/30 of the distance S_o between pilasters or ice structure columns, the ice ring beam can be considered as the supporting point of fixed hinges of the wall between pilasters or the wall between ice structure columns, and element height H shall be the distance between adjacent fixed hinges; when the width of ice ring beam is not allowed to be increased, the height of the ice ring beam can be increased according to the constant stiffness principle of the wall surface.

4. 4. 15 The ice masonry construction shall meet the following requirements:

1 When the overall height of the battened hollow ice wall exceeds the allowed ratio of height to thickness, the ice masonry construction shall meet the following requirements:

1) The thickness of the single wall shall not be less than 250mm;

2) The battened hollow ice wall shall be connected with ice blocks and two binding ice blocks between which horizontal steel plates (3mm thick) shall be set. The width of connecting ice blocks shall not be less than the sum of the two single walls. The overlapped part of each block shall not be less than 200mm; the two kinds of ice blocks above shall be set alternately along the height of the battened hollow ice wall. The space between the blocks shall not be greater than half of the permissible ratio of height to thickness of a single wall.

2 The sectional dimension of the independent supporting hollow ice column shall not be less than 600mm×600mm, with wall thickness not less than 200mm; the sectional dimension of the solid ice column shall not be less than 400mm×400mm.

3 When the height of the independent ice column exceeds 15m, the reinforcement set in the ice column shall meet the following requirements:

    1) The ratio of vertical reinforcement shall be no less than 0.2%, and the reinforcement shall be no less than 8Φ16. Ribbed steel bars shall be used;

    2) Vertical reinforcement shall be connected by lapping, mechanical connections or welding; when lapping is adopted, the overlapping length shall not be less than $60d$ (d is the larger value of the diameter of the steel bars involved in the overlapping), and not less than 1200mm. The anchorage length shall not be less than $80d$, and not less than 1500mm;

    3) The diameter of hoop reinforcement shall not be less than Φ12, and the spacing between shall be no more than a triple ice block, and no more than 600mm.

4 Ice masonries shall be laid layer by layer in a staggered manner, and the overlapped length of the stagger joint shall not be less than 120mm.

5 The expansion joints of ice masonries shall meet the following requirements:

    1) The maximum spacing between expansion joints shall be no more than 30m;

    2) The width of expansion joint shall not be less than 20mm. with cold-resistance and moistureproof elastic material filled through the joint, with no debris in the joint.

6 Steel plate mesh or transparent partition shall be set at the top of the entrance when there are people or vehicles passing through the arch entrance of an ice masonry and the entrance is 3m wide.

7 For a large-volume ice building or ice landscape with crushed ice filled inside, when the height of the external ice wall is greater or equal to 6m, the laying thickness of the ice wall shall not be less than 900mm; when the height of the external ice wall is less than 6m, the laying thickness of the ice wall shall not be less than 600mm and shall meet the requirements for the ratio of height to thickness of the ice wall.

4.4.16 In an area involving seismic fortification, for an ice landscape building that has a height of more than 12m or whose number of stories are greater than four, corresponding seismic construction measures shall be taken based on the possible destruction caused by earthquakes.

4.4.17 Lintel setting shall meet the following requirements:

1 The flat arch entrance of an ice masonry shall not be wider than 3m. The section steel lintel shall be selected based on Table 4.4.17-1.

Table 4.4.17-1 Table of channel steel and angle steel lintel selection

Width of Ice Masonry Portal L_n(mm)	Section Steel Type	Section Steel Spacing (mm)	Section Steel Standard and Quantity
$L_n<1000$	Channel steel	500	2[8
	Angle steel	500	2L50×5
$1000 \leqslant L_n<2000$	Channel steel	500	2[10
	Angle steel	500	2L75×6

Table 4.4.17-1(continued)

Width of Ice Masonry Portal L_n(mm)	Section Steel Type	Section Steel Spacing (mm)	Section Steel Standard and Quantity
2000≤L_n≤3000	Channel steel	500	2[12
	Angle steel	500	2L110×8

Notes: 1 The ice masonry over the section steel lintel shall be laid layer by layer in a staggered manner, and the overlapped length of the stagger joint shall be half of the ice block. When there is additional load on ice masonry over the lintel, section steel standard shall be determined through calculation;

    2 The support length of section steel lintel shall not be less than 300mm.

2 When an ice arch lintel with ice block is used, the size and rise of the ice lintel shall follow Table 4.4.17-2.

Table 4.4.17-2 Size and rise of ice arch

Width of Ice Portal L_n(mm)	Height of Wedgy Ice Arch d (mm)	Rise f_o(mm)
L_n≤3000	d≤300	f_o≤1500
3000<L_n≤6000	300<d≤600	1500<f_o≤3000
6000<L_n≤9000	600<d≤900	3000<f_o≤4500

Note: 1 The wedged ice arch is an arched portal. When the height of the ice arch is over 550mm, it shall be laid in two layers and the height of it is the sum of the two layers of wedged ice blocks;

    2 The ice masonry upward the portal shall be built layer by layer in a staggered manner, and the overlapped length of the stagger joint shall be no less than half of the ice block length;

    3 The ice arch height shall not be less than 1/10 the portal width and the ice arch rise shall not be less than 1/2 the portal width.

3 Considering the horizontal cross-section bearing capacity of ice masonry arch foot pedestal, shearing and slipping resistance calculation shall be conducted according to the pushing force of the arch foot. Considering the decreasing bearing capacity because of thawing, corresponding construction measures shall be taken.

4.4.18 When the cantilever beam length exceeds 0.6m, section steel cantilever beam structural treatment shall be used based on the cantilever structure.

4.4.19 Anti-overturning verification and calculation of the section steel cantilever beam in the ice masonry wall shall be conducted according to the current national standard *Code for Design of Masonry Structure* GB 50003.

4.4.20 When the height of an ice landscape building is more than 12m, or the number of stories is greater than four stories, rigid connections or floors shall be designed at the elevation of the ring beam. The main load-bearing structure of floors and roofs shall be fabricated steel structures equipped with a purline system, and section steel can be used for the spandrel girder.

4.5 Structural Design of Snow Construction

4.5.1 The structure elements of snow masonries shall be designed based on the ultimate state of the bearing capacity, and meet the requirements under normal conditions.

4.5.2 When the structure elements of snow masonries are designed based on the ultimate state of the bearing capacity, calculation shall be conducted based on the most unfavorable combination of the following formulas:

$$\gamma_0 (1.2S_{Gk} + 1.4\gamma_L S_{Q1k} + \gamma_L \sum_{i=2}^{n} \gamma_{Qi}\psi_{ci}S_{Qik}) \leqslant R_d \qquad (4.5.2\text{-}1)$$

$$\gamma_0 (1.35S_{Gk} + 1.4\gamma_L \sum_{i=1}^{n} \psi_{ci}S_{Qik}) \leqslant R_d \qquad (4.5.2\text{-}2)$$

Where: γ_0 ——structure importance coefficient, 1.0 taken;

γ_L —— for variable loads, considering service life adjustment coefficient in structure design, only limited to floors and roofing, 0.9 taken;

S_{Gk} ——effect of permanent load standard value;

S_{Q1k} ——effect of a variable load standard value that plays a controlling role in basic combinations;

S_{Qik} ——effect of the ith variable load standard value;

R_d ——design value of resisting force of structure elements;

γ_{Qi} ——subentry coefficient of the ith variable load, 1.4 taken;

ψ_{ci} ——coefficient of the combination value of the ith variable load, 0.7 taken.

4.5.3 The bearing capacity of structure elements of snow masonries shall comply with the following provisions:

1 The bearing capacity of structure elements of ice masonries shall be calculated based on the ice masonry strength value at $-10°C$ in the temperature grading;

2 During the calculation of dead weight of ice masonries, conversion to gravity density (kN/m³) shallbe conducted according to the values in Table 3.2.1;

3 The values of dead weight of non-snow masonries structure elements and the acting load shall be taken according to the relevant provisions of the current national standard *Load Code for the Design of Building Structure* GB 50009.

4.5.4 The foundation design of snow masonry building shall meet the following requirements:

1 For the snow masonry building whose height is over 10m and the short edge exceeds 6m, foundation design shall be undertaken. The bearing capacity of the foundation shall be calculated based the strength of non-frozen soil, with consideration given to the frozen expansion of the soil around it, with corresponding anti-frozen expansion measures taken;

2 For the snow masonry building whose height is more than 10m, when the natural foundation design conditions cannot be met, the use of frozen soil by pouring water or using other reinforcing methods can be adopted. The bearing capacity of the treated foundation shall meet the design requirements. As for frozen soil foundation, underlying layer calculation and foundation frozen expansion stability calculation shall be conducted.

4.5.5 For a snow masonry building whose building height is less than 10m, frozen soil ground base can be adopted which can be obtained by watering and freezing the watered soil. When the thickness of frozen soil is more than 400mm, 400mm shall be taken; when the thickness is less than 400mm, the actual thickness of frozen soil shall be taken. The bearing capacity of frozen soil foundation shall be determined with the in-situ test. The underlying layer calculation and stability calculation of frozen soil foundation shall be conducted.

4.5.6 For the snow masonry building, a static force calculation scheme for static force calculation can be determined in accordance with the current national standard *Code for Design of Masonry Structures* GB 50003 and design can be conducted according to a rigid scheme.

4.5.7 The bearing capacity of elements in compression shall use the following formula:

$$N \leqslant \varphi f A \qquad (4.5.7)$$

Where: N——design value of axial pressure;

φ——The influence coefficient of β (the ratio of height to thickness) and e (eccentricity of axial force) to elements in compression shall follow Appendix B of he standard; The ratio β of height to thickness shall be calculated with 4.5.12-1 and 4.5.12-2 of this regulation; When e is calculated based on endogen force design value, it shall be no less than 60% of the distance from gravity center to the edge of axial eccentric section. The compressive strength of snow masonry shall follow Table 3.2.2 of the standard;

f——the design value of compressive strength of snow masonry to be taken according to the provisions in Table 3.2.2 of the standard;

A——area of section. Snow masonry shall be calculated based on the net cross section; the flange widths of the wall with pilasters and the wall with ice structure columns shall follow Items 1 and 2 of 4.5.12-2 of the standard respectively. The net section length shall be taken for the walls between pilasters or ice structure columns.

4.5.8 The bearing capacity of local compression shall use the following formula:

$$N_l \leqslant 1.2 f A_l \qquad (4.5.8)$$

Where: N_l——design value of axial force in a local compression area;

f——an snow masonry's design value of compressive strength, value taken based on the provisions in Table 3.2.2 of the standard;

A_l——area in compression.

4.5.9 The bearing capacity of the element in axial tension shall use the following formula:

$$N_t \leqslant f_t A \qquad (4.5.9)$$

Where: N_t——design value of axial tension;

f_t——design value of tensile strength of snow masonry, value taken based on the provisions in Table 3.2.4 of the standard;

A——section area, calculated based on net cross section of the snow masonry.

4.5.10 The bearing capacity of the element subject to shearing force shall use the following formula:

$$V \leqslant f_v A \qquad (4.5.10)$$

Where: V——the design value of shearing force;

f_v——design value of the shearing strength of snow masonry, value taken based on the provisions in Table 3.2.5 of the standard;

A——section area, calculated based on the net cross section of the snow masonry.

4.5.11 The bearing capacity of the element subjected to bending force shall follow this formula:

$$M \leqslant f_w W \qquad (4.5.11)$$

Where: M——design value of cross section bending moment;

f_w——design value of the bending strength of snow masonry, which can adopt the design value of bending strength, value taken based on the provisions in Table 3.2.3 of the standard;

W——cross section resistance moment.

4.5.12 The ratio of height to thickness of walls and columns shall be in accordance with the following provisions:

1 The ratio of height to thickness of snow walls and columns shall follow this formula:

$$\beta = \frac{H_0}{h} \leqslant [\beta] \qquad (4.5.12\text{-}1)$$

Where: H_0——calculated height of walls and columns, see Table 4.5.12-1;

h——thickness of wall or length of the shorter edge of the rectangle column;

$[\beta]$——the permissible ratio of height to thickness of walls and columns, see Table 4.5.12-2.

Table 4.5.12-1 Calculated height of walls and columns H_0

Snow masonry Building and Element Category	Floor and Roof Category	Spacing between Transverse Walls S(m)	Wall with Pilasters or Wall with Ice Structure Columns or Wall Connected with Others		
			$S>2H$	$2H \geqslant S > H$	$S \leqslant H$
Rigid scheme for snow masonry buildings	Fabricated light-duty floor and roof with purlin	$S<20$	1.0H	0.4S+0.2H	0.6S
	Wooden or light steel roof of tile roofing	$S<16$			
Non-rigid scheme for snow masonry buildings	Fabricated light-duty floor and roof with purlin	$S \geqslant 20$	1.5H		
	Wooden or light steel roof of tile roofing	$S \geqslant 16$			
Upper end is free			2.0H		

Note: 1 In the base course, H is the distance from top floor (or the level of supporting points) to bottom bearing of the element; on other floors, H is the distance between floors or other horizontal supporting points;

2 When the upper end of element is free, the height H equals the length of the element;

3 For the gable without pilasters or ice structure column, H is to be the height between floors plus a half of the gable; for the gable with pilasters or ice structure column in snow masonry, H is the height of the gable wall at the place of the pilaster or ice column;

4 For the three-side supporting walls without a cap, H is the distance from the upper free distance to bottom bearing point, there shall be an ice ring beam, pilaster or ice structure column.

Table 4.5.12-2 Permissible ratio of height to thickness of walls and columns $[\beta]$

Member	Snow Wall	Snow Column
Main load-carrying member	8	6
Minor load-carrying member	10	8

2 The ratio of height to thickness of wall with pilasters and wall with ice structure column shall be verified and calculated based on the following formula:

$$\beta = \frac{H_0}{h'} \leqslant [\beta] \qquad (4.5.12\text{-}2)$$

Where: H_0——calculated height of wall with pilaster, wall with ice structure column in snow masonry or wall between pilasters, wall between ice structure columns, see table 4.5.12-1 and Table 4.5.12, Item 2, Subitem 3 respectively;

h'——the converting thickness of the wall with pilasters and wall with ice structure columns, see Table 4.5.12, Item 2, Subitems 1 and 2. The thickness of wall between pilasters and wall between ice structure columns, shall be the thickness of the wall itself;

$[\beta]$——the permissible ratio of height to thickness of wall and column, see Table 4.5.12-2.

1) The converting thickness of the column wall with pilaster shall be 3.5 times the gyration radius of the section. For the strip foundation of wall with pilaster, flange width of the section of the wall with pilasters can be the distance between adjacent pilasters; For single story snow masonry buildings, flange width of the section can be the width of the pilaster plus 2/3 the height of wall, but it shall not exceed the width of the wall between windows, or the distance between two adjacent pilasters. For multi-story snow masonry buildings that have walls with doors and windows, the width of the flange of the section of the wall with pilasters can be the width of the snow wall; for walls without doors and windows, the width of the flange wall on each side can be 1/3 of pilaster height and shall not exceed the distance between two adjacent pilasters;

2) The flange width of the section of wall with ice structure columns in the snow masonry can be the distance between two adjacent ice structure columns. The converting thickness shall be 1.05 times the thickness of the wall;

3) During the verification and calculation of the ratio of height to thickness of wall between pilasters or between ice structure columns, spacing S between transverse walls shall be the distance between pilasters or structure columns; calculated height H_0 of the wall with pilasters or ice structure columns that has an ice ring beam shall follow Table 4.5.12-1, but element height H shall be defined according to the following rules: when the ice ring beam b is not less than 1/30 of the distance S_o between pilasters or ice structure columns, the ice ring beam can be considered as the supporting point of fixed hinges of the wall between pilasters or the wall between ice structure columns, and element height H shall be the distance between adjacent fixed hinges; when the width of ice ring beam is not allowed to be increased, the height of the ice ring beam can be increased according to the constant stiffness principle of the wall surface.

4.5.13 The snow masonry construction shall meet the following requirements:

1 For the snow wall whose height is less than 6m, its thickness shall not be less than 800mm. For the snow wall whose height is more than 6m but less than 10m, its thickness shall not be less than 1000mm; the section size of the independent snow column section shall not be less than 1200mm×1200mm;

2 For the snow wall and the independent snow column over 10m in height, it shall be reinforced with bamboo, wood or steel structures;

3 The arch portal with a span wider than 2m, the portal for people and vehicle flows, shall be reinforced with bamboo, wood or steel structures outside the snow masonry.

4.5.14 In the area where seismic resistance is needed, the snow masonry building whose construction height is higher than 9m or that has more than three stories, shall have corresponding seismic construction measures to avoid the possible damage caused by earthquakes.

4.5.15 Lintel setting shall meet the following requirements:

1 The flat arch portal of snow masonry shall not be wider than 3m, with section steel lintel to be selected based on Table 4.5.15-1.

Table 4.5.15-1 Table of channel steel and angle steel lintel

Width of Snow Masonry Portal L_n (mm)	Section Steel Type	Section Steel Spacing (mm)	Section Steel Standard and Quantity
$L_n < 1000$	Channel steel	500	2 [8
	Angle steel	500	2 L50×5
$1000 \leqslant L_n < 2000$	Channel steel	500	2 [10
	Angle steel	500	2 L75×6
$2000 \leqslant L_n \leqslant 3000$	Channel steel	500	2 [12
	Angle steel	500	2 L110×8

Note: 1 The snow masonry over the section steel lintel shall be built layer by layer in a staggered manner, and the overlapped length of the stagger joint shall not be less than a half of the snow block. In the case of additional load on the snow masonry over the section steel lintel, section steel standard shall be determined through calculation;

2 The support length of section steel lintel shall not be less than 400mm.

2 When a snow arch lintel with snow block is used, the size and the height of the snow lintel shall be based on Table 4.5.15-2.

Table 4.5.15-2 Size and rise of snow arch

Width of Snow Portal L_n (mm)	Height of Wedgy Snow Arch d (mm)	Rise f_o (mm)
$L_n \leqslant 3000$	$d \leqslant 500$	$f_o \leqslant 1500$
$3000 < L_n \leqslant 6000$	$500 < d \leqslant 800$	$1500 < f_o \leqslant 3000$
$6000 < L_n \leqslant 9000$	$800 < d \leqslant 1100$	$3000 < f_o \leqslant 4500$

Note: 1 The wedge snow arch is for an arched portal. When the height of the snow arch is over 550mm, it shall be built into two layers and the height of the wedgy snow arch is the sum of the two layers of wedge snow blocks;

2 The snow masonry over the structural steel lintel shall be built layer by layer, and the overlapped length of the stagger joint shall not be less than a half of the snow block;

3 The snow arch height shall not be less than 1/10 the portal width and the snow arch rise shall not be less than 1/2 the portal width.

3 Considering the horizontal cross-section bearing capacity of a snow masonry arch foot pedestal, shearing and slipping resistance calculation shall be conducted according to the pushing force of the arch foot. Considering the decrease of the bearing capacity because of thawing, corresponding construction measures shall be taken.

4.5.16 When the cantilever beam length exceeds 0.4m, section steel cantilever beam shall be used. Anti-overturning verification and calculation of the section steel cantilever beam in the snow masonry wall shall follow the requirements of *Code for Design of Masonry Structure* GB 50003.

4.5.17 When the height of a snow landscape building is greater than 9m, or the number of stories of the snow landscape building is over three, rigid connections or floors shall be designed at the elevation of the ring beam. The main load-bearing structure of floors and roofs shall be fabricated steel structures equipped with a purlin system, and section steel can be used for the spandrel girder.

4.6 Illumination Design of Snow and Ice Landscapes

4.6.1 The illumination design of ice and snow landscapes shall follow the relevant provisions of existing national standard *Standard for Lighting Design of Buildings* GB 50034 and meet the requirements for electric power of civil buildings and urban night landscape lighting.

4.6.2 Overall lighting design and monomer lighting design shall be conducted for landscape lighting in the scenic area and meet the following provisions:

    1 Both inside and outside lighting facilities shall be designed for ice and snow landscape lighting;

    2 The lights' color and illumination and change frequency shall be rationally allocated according to the theme;

    3 The arrangement of lighting equipment shall comply with the general lighting design and monomer lighting design requirement, and a reasonable position shall be determined. The luminance illumination, and the lights' color and shadow shall comply with the lighting design effect requirements;

    4 The lighting equipment shall be environmentally friendly, energy saving and highly efficient;

    5 Make sure that the outdoor ancillary facilities such as lighting equipment, and frameworks can work properly in cold weather.

4.6.3 The illumination quality of ice and snow landscapes shall follow the regulations below:

    1 The color temperature of the illuminating light source for ice and snow landscapes shall follow the provisions in Table 4.6.3-1.

Table 4.6.3-1 Color temperature of illuminating light source for ice and snow landscape buildings

Light Source Color Classification	Color Temperature (K)	Color Characteristics	Applicable Landscape
I	<3300	warm	Classical, European-style ice architecture and commercial facilities
II	3300~5300	middle	Ice sculpture works and advertising
III	>5300	cold	Ice and snow sculpture, recreation activities

    2 The color of lighting source shall match the theme of the ice lanterns and snow sculptures.

    3 For ice and snow landscape lighting, the quality of lighting direct glare Level (*UGR* value) shall follow Table 4.6.3-2.

Table 4.6.3-2 For ice and snow landscape lighting the quality of lighting direct glare level(*VGR*-value)

Value of *UGR*	Description of Corresponding Degree of Glare	Examples of Vision Requirements and Places
<13	No glare	—
13~16	Noticeable	Ice sculpture
17~19	Drawing attention	Ice sculpture
20~22	Causingminor discomfort	Snow sculpture
23~25	Uncomfortable	Snow sculpture and scenic area illumination
26~28	Very uncomfortable	—

4 In places where ice and snow landscape lighting is used, if the (*UGR*) is greater than 25, the following measures against glare shall be taken:

    1) For large scale snow sculptures and ice buildings, lighting equipment shall not be set at the interfering space or in an area where it may form specular reflection to the eyes;

    2) For small ice and snow landscape architecture and artistic ice sculpture illumination, light distribution or indirect measure of lighting along line-of-sight direction could be adopted, with lighting equipment that has a big area of light-emitting surface, low brightness and proliferation of good optical performance.

4.6.4 Illumination level of ice and snow landscape buildings shall meet the following requirements:

  1 The visual working illumination range value shall be preferably selected based on Table 4.6.4.

Table 4.6.4 Visual working illumination range value

Nature of Visual Working	Illumination Range(lx)	Region or Type of Activity	Examples of Applicable Places
Simple visual working	30~75	Simple identification of materials characterization	Places for entertainment Activities
General visual working	100~200	Landscape inside lighting, commercial workplace, etc.	Inside the ice sculpture and ice landscape buildings
	200~500	Projector lighting outside the landscape	Ice sculpture sketch, small-scale snow sculpture
	500~750	Large-scale ice and snow landscape buildings, landscape display areas, important viewing places	Important landscape area such as symbolic landscapes, stage performances, etc.

  2 Landscape illumination of ice and snow landscape buildings may use the following classification (lx): 20, 50, 100, 200, 300, 500 and 750.

  3 The performance lighting in the performance area shall be equipped with a professional illumination system based on the performance requirements.

  4 The roads in a scenic area shall be equipped with functional lighting equipment, and a single-color source with a favorable illumination level of 20 lx~50 lx is preferred. Indirect lighting can also be applied such as light boxes, advertising lights or lights embedded in the ground.

  5 The scenic area surrounding lightning effects shall involve rational design the combination of bright and dark effects, color change and rational control of combination of points, lines and planes of the light source. When lights with special performance such as lasers are used, program control shall be provided, and the illumination degree in the central scenic area and main landscapes shall be higher than that in the other scenic areas.

  6 The lighting design shall include preparing illumination distribution, scenic area and large-scale landscape light color effect sketches, lighting change program design and dynamic demonstration scheme.

4.6.5 Light source and lighting equipment for ice and snow landscape architecture shall meet the following requirements:

1 Light source and lighting equipment in landscape areas shall be selected and configured for a good visual effect, reasonable distribution, appropriate light illumination, prominent colors and appropriate exchange.

2 Light sources shall be chosen considering factors such as scenic area environment, light effect, color rendering property and durability.

3 The light sources inside the icescape shall be energy-saving, environmentally friendly, waterproof, with low heat generation amount (LED, etc.).

4 LED can be applied in designated areas for advertising, information release, guide maps and big screens.

5 Electroluminescent panel can be used in the guiding signs as auxiliary lighting.

6 Landscape illuminating equipment shall be selected with the qualities of low heat-production, safety, energy conservation, environmental friendliness and durability, and shall also be suitable for use at low temperature. In addition, the following requirements shall be met:

    1) The lamps inside the icescape and clearance light which contact the icescape shall be high-bright and luminescent;

    2) Those lamps inside the icescape which cannot be dismantled shall be low-cost, less-polluting and durable;

    3) The outdoor lamps shall be water resistant, moisture resistant and easy to replace.

7 Durable spotlight, floodlight and upper-air signal lamp (luminaries with durable light source) shall be used when the lamps cannot be easily overhauled and maintained.

8 Bright gas discharge lamps shall be used for functional lighting in the scenic area and square.

4.6.6 The lighting of ice and snow landscape buildings shall adopt energy-saving measures, the following requirements shall be met:

1 Light sources which are economical, eco-friendly and energy saving shall be chosen.

2 Straight tubular fluorescent lamps shall use energy-conserving electronic ballasts which can work normally at low temperature. When electronic ballasts are used, the energy consumption of the fluorescent lamps shall meet the current national standard *Limited Values of Energy Efficiency and Evaluating Values of Energy Conservation of Ballasts for Tubular Fluorescent Lamps* GB 17896.

3 Lighting control modes shall be properly selected to meet the lighting design requirements:

    1) The light may be controlled separately or in a multipoint way;

    2) Illumination in public shall have a centralized control, or clock-control;

    3) A light reducing scheme for different time periods can be provided;

    4) The voltage can be properly reduced when high-efficiency lanterns are turned on.

4 Such energy-saving management measures as infrared sensor controls, time switches and intelligent lighting control systems shall be used to satisfy the scenic area's illumination function requirements in the scenic area.

5 Lamps taking advantage of solar power and wind energy can be used for the illumination on the square, roads and courtyard.

4.6.7 The lighting power supply system for ice and snow landscape buildings shall meet the following requirements:

1 The load level and power supply solutions shall be properly designed.

2 The major lighting power load supply system shall adopt a duplicate supply and double-loop supply. Each level of the systems shall have its own power control.

3 The load balance of a three-phase lighting line shall be kept. The maximum phase load current shall not be greater than 115% of a three-phase load average. The minimum phase load current shall not be less than 85% of the three-phase load average.

4 In important areas, for the end load distribution box, power shall be supplied by the power source which switches automatically. For the heavy load, power can be supplied by two particular circuits with a 50% load for each.

5 In the branch circuit of illumination, three-phase low-voltage circuit breakers shall not be applied to control and protect three-phase branch circuits.

6 The electric current of each single-phase branch circuit in the lighting system shall not exceed 16A, the number of light sources shall not be greater than 50. The electric current of each single-phase branch circuit in the combined lanterns of a large ice building shall not exceed 25A, and the number of light sources shall not be greater than 120.

7 Neutral conductors and phase conductors shall follow the same standard when gas discharged lamps are used for lighting lines.

8 The same light or the adjacent tubes of different light (light source) can be connected to the lines of different phases when inductance ballasts gas discharge lights are used.

9 The overall power supply solution shall consider the scenic area planning and monomer ice and snow landscape building lighting design to calculate electrical load, determine power supply solutions and complete power distribution system design and shall meet the following provisions:

    1) When a constant power supply is used, power distribution equipment and power supply lines shall be fixed facilities. Power supply lines shall be buried directly underground; protective measures shall be taken where such lines run through roads or heavy-duty vehicles passes;

    2) When temporary power supply is adopted, power distribution lines shall be laid in metal troughs either visibly or invisibly.

10 On-duty personnel in charge of lighting are to be stationed in the landscape area.

11 Electric equipment and circuit breakers (including miniature circuit breakers) shall be able to work properly below $-30°C$.

12 For outdoor branch circuits, RCD shall be installed.

13 A short-circuit protector, overload protector and overvoltage and undervoltage protectors shall be installed on power distribution lines.

14 The protection grade for an outdoor power distribution cabinets or power distribution boxes shall be no less than IP33.

4.6.8 The lighting design of ice and snow landscape buildings shall meet the following requirements:

1 The main lighting source shall be appropriate to the theme of the icescape architecture. The light shall be changeful. The lighting system of large scale ice landscapes shall be controlled with a program-control system. Art icescapes can adopt projection light, of which the color, illumination, disposition of lighting and the type of lanterns shall be in accordance with the theme and its artistic expression.

 2 Light source selection shall meet the following provisions:

  1) The chromatic aberration of the rendering light color in the ice body shall not be too small. It is preferred to choose primary colors, such as white, yellow, red, blue and green, as sheme;

  2) Tubular fluorescent lamps or moldable LED lamps shall be used inside the icescape;

  3) The clearance lights of the icescape can be neon light, moldable LED lamps, optical fiber lamps, flash lamps or strobe lamps, etc;

  4) Gas-light shall be the main light source of the icescape architecture, luminaries and spotlights are preferred;

  5) Halogen lamps, flashlights, spotlights and some other warm light can be used to show the effect of local lighting of the ice sculpture and large icescape buildings. These lamps shall be energy-saving and compact.

 3 T8 tubular fluorescent lamps, or T5 small size tubular fluorescent lamps combined with a color filter plate shall be preferably used as the inner light source of the ice architecture. Tricolor tubular fluorescent lamps, energy-saving compact lights and LED can also be used.

 4 For the ice architecture and anaglyptic icescapes that are more than 3m high and 1.5m wide and thick, the main light source shall be embedded lamps. Floating light for complementary color can be used locally. Program control shall be designed for the light source to make the light sparkling, flowing and colorful. The distance between built-in lights and the surface ice is determined by the transparency of the ice. Usually, the distance is between 150mm and 350mm.

 5 The cast lights shall be used as the light source for the artistic icescape. The distance between lamps and landscapes shall be greater than 1.5m. The lamps shall be set snugly and at a certain angle with the icescape. The type, color, intensity and distance of the main light source and the assisting light source shall meet the needs of manifestation effect. The lamps shall be installed on lamp hangers whose height from the ground shall not be less than 0.5m.

 6 Incandescent lamps and halogen lamps can be used as the main light source of the local icescape, assisted with spot lights.

 7 Large scale starry light can be used in large areas such as ice galleries and bushes.

 8 The new LED light source embedded in the icescape shall preferably meet the following requirements:

  1) Considering the directionality of the light source, the light distribution in the ice masonry shall be uniform;

  2) The selected light types, accessory standard and models shall be unified and universal;

  3) It shall be preferred not to use equipment with large heat dissipation amount for the light source and accessories in the ice masonry;

  4) Recycling measures shall be taken for reusable lamps and electrical materials.

4.6.9 The lighting design of snow landscapes shall meet the following requirements:

 1 Metal halide lamps and high-voltage sodium lamps can be used for snow sculptures. The distance between the lamps and snow landscapes shall be greater than 2.0m;

 2 Lighting devices shall be installed on a lamp holder and shall be as least 0.5m above the ground;

3 Light color and its illumination shall be appropriate to the design theme. Key light, side light and back light shall be designed to achieve the corresponding effects;

4 The key light and other lights for large icescapes shall be distinguishable. The power of the flood lamp directly illuminating the snow landscape shall be less than 400W;

5 Small floodlights shall be preferably used. Lamps and accompanying stands shall be preferably painted white.

4.6.10 The ground connection for low-voltage distribution system in ice and snow scenic areas shall meet the following requirements:

1 The grounding form of the low-voltage distribution system in ice and snow scenic areas shall adopt TT or TN-S system.

2 If a TT system is adopted, every power distribution box shall be equipped with an earth electrode and the operating characteristic of the earth fault protection shall comply with the provisions in the formula below:

$$R_A \times I_a \leqslant 50 \qquad (4.6.10)$$

Where: R_A——the sum of the resistance of the earth electrode and the resistance of e protective conductor of the exposed conductive part (Ω)

I_a——the operating current cutting off the fault circuit to protect appliance (A)

3 If an over-current protector is adopted, the operating current of the fault circuit of inverse time over-current protector (I_a) shall cut off the current within 5 seconds. The operating current (I_a) which adopts the fault circuit of instantaneous characteristics of the over-current protector, shall be the minimum instantaneous current ensuring instantaneous action. If a residual current operating protector is adopted, it shall be the rated residual operating current.

4 If a TN-S system is adopted and the protective electric conductor (PE) exceeds 50m, then the grounding shall be repeated. If the line is too long, the exposed conductive parts at the end of the distribution box, as well as the external conductive parts, are supposed to be local or auxiliary equipotential bonding.

4.6.11 Grounding methods of distribution distribution lines, equipotential bonding and protection shall conform to the current national standard *Code for Design of Low Voltage Electrical Installation* GB 50054.

4.7 Intelligentization Design

4.7.1 Broadcasting system and sound system should be devised in the scenic area.

4.7.2 The sounds and light music shows of separate performing areas should not disturb other areas in the scenic area.

4.7.3 Wireless explanation system and wireless mobile terminal which provide electronic maps, attractions commentary, travel route query, traffic information, scene navigation, commercial navigation, location services and other related services to tourists should be put to use in the scenic area.

4.7.4 The monitoring system for continuous monitoring of the area, and the automatic alarm system in important areas should be set up in the scenic area.

5 Construction of Ice and Snow Landscape Buildings

5.1 General Requirements

5.1.1 Before the construction, the operating organization should assemble all of those involved in design, construction and supervision to make a joint checkup of the drawings and technical disclosure.

5.1.2 Those involved in building the environment should compile organizational designs for the construction, and work out the schemes accordingly. Bearing capacity and stability of the bracing structure should be calculated and supervised, determining the technical measures for working aloft; construction survey; machine selection; embedding of profiled bars; installation of building blocks; ice and snow incision and transportation.

5.1.3 **For the height of the ice building more than 30m or that of the snow building more than 20m, sedimentation and deformation observation during the construction should be taken.**

5.1.4 On-spot inspection of the materials and equipment involving structural safety and functions of use should also be undertaken.

5.1.5 After the completion of the main building construction of ice and snow landscape, the surface of ice landscape should be cleaned and the surface of snow masonry should be polished so as to achieve the effects of transparence for ice masonry, and bright and clean surface for snow masonry.

5.2 Construction Survey

5.2.1 The construction of ice and snow landscape building should meet the design requirements to generally set off lines of the site. The location of the individual landscape and its pile point should be well-protected after it has been properly determined through inspection.

5.2.2 The profile should be measured on the basis of the set stakes or control point of the ice and snow landscape. After its closed calibration gives a satisfactory result, the detail axis and relevant borderline should be in accordance with the permissible deviation shown in Table 5.2.2

Table 5.2.2 Permissible deviation of the detail axis

Item		Permissible deviation
Detail Axis		±10mm
Elevation	Floor Height	±15mm
	Total Height	±30mm
Total Height Verticality (m)	$H \leqslant 15$	±20mm
	$H > 15$	The minor of $H/750 \pm 50$mm
Outline Side Length (m)	$L(B) \leqslant 30$	±20mm
	$L(B) > 30$	±30mm
Diagonal Line (m)	$L(B) \leqslant 30$	±30mm
	$L(B) > 30$	±40mm
Axes Angle (")	$L(B) \leqslant 30$	±20"
	$L(B) > 30$	±30"

5.3　Ice-collecting and Snow-making

5.3.1　The collection of natural ice should follow the following regulations:

1　The suitable temperature for natural ice collection is below $-10°C$.

2　Only when the thickness of natural ice is not less than 200mm, and the ice material meets the following conditions, can the ice be collected:

1) The strength can meet the demand of the design;

2) A good transmission of light, no obvious air bubbles, silt, sundries as well as cracks and no possibilities of the ice breaking away (abruption) are required.

3　Under natural conditions, rough ice should be kept for more than 12 hours before use.

4　Rough ice should match the following standard: the length is 1000mm, the width is 700mm and the thickness is more than 200mm, or 1300mm long, 1200mm wide and not less than 300mm thick. The ice sculpture should adopt the whole rough ice of 2000mm long, 1200mm wide and not less than 400mm thick.

For the masonry, the ice block should have a length of 600mm, a width of 300mm and a thickness of not less than 200mm.

5.3.2　Rough ice should be divided by rack saw, and processed into ice masonry as required.

5.3.3　Artificial ice making should follow the following regulations:

1　The temperature of the environment should be below $-10°C$;

2　Measures should be taken to make the ice transparent when making transparent artificial ice;

3　When making colored ice, the colored dye should be easy to dissolve in water, pollution-free, good in suspension and light transmission. It should meet the requirements of environmental protection. The degree of hue saturation should conform to those of the design;

4　The dimensions of the artificial ice could be 600mm×300mm×200mm.

5.3.4　The fabrication of artificial snow should follow the following regulations:

1　The temperature of the environment should be below $-10°C$;

2　When making snow on a large scale, the water supply should be sufficient and its quality should meet the standard of the snowmaker;

3　When making snow indoors, a snowmaker with a high degree of atomization and thinner spray nozzle is preferred. Snow can also be made with crushed ice using big ice crushers.

5.3.5　Large snow blocks can be made in the following ways:

1　When adopting the method of snow accumulation, templates can be used according to the design requirements to form geometries, which are then filled with snow and compacted layer by layer;

2　When adopting the method of snow building, the snow blocks whose strength meets requirements should be employed to form larger snow blocks. The joint and geometric dimension of the snow block should be neat.

5.4　Foundation Construction of Ice Building

5.4.1　Before the construction, level and smooth the ground base, ensuring it is entirely frozen after water flooding, and then constructs the upper masonry.

5.4.2 If the slope of the ground base is less than 1‰ and the height difference is less than 100mm, level and smooth it by watering and freezing entirely. If more than 1‰ or the height difference is more than 100mm, then level and smooth it by ice masonry.

5.4.3 The bearing wall and columns should be built on the ground base. It is forbidden to base it on the crushed ice, snow and loose soil layers.

5.4.4 The base construction of ice building should meet the following regulations:

    1 Adopt the method of combining and piling up layer-separated ice blocks, the upper and lower part should lap off the set joints. The lapping length should be half of that of the ice block, and should not be less than 200mm. The method of making masonry on the surrounding sides and filling in the middle should be avoided.

    2 The masonry height of every layer of ice should be horizontally identical. The horizontal and vertical seams of ice masonry should be less than 2mm wide and be frozen by pouring water. The ratio of frozen area of ice seams should be more than 80%.

    3 The ice building which is designed to be hollow or filled with crushed ice should have a foundation bed, the height of which should be 1/10 that of the ice building and not less than 1m.

5.5 Construction of Ice Masonry

5.5.1 The exterior of the ice landscape building should adopt ice blocks which has high transparency with no incisions or cracks.

5.5.2 The freezing water between the ice blocks should be natural water or tap water.

5.5.3 The ice landscape should be constructed by combining and piling up ice blocks layer by layer.

5.5.4 The temperature for poured water during the construction should be 0℃, and special infusing tools should be adopted to perfuse ice crack, and the frozen rate should not be less than 80%.

5.5.5 During the construction, monitoring the temperature of ice masonry should be carried out. When the temperature is higher than the design temperature, or the masonry water cannot be frozen, construction should be suspended and methods should be adopted to protect the ice landscape, such as using shading and wind-proofing, etc.

5.5.6 The size of ice blocks should be determined by the thickness of the ice masonry (wall) and the size of the ice material. Every masonry side should be leveled up. And the permissible deviation of the ice height is ±5mm. The allowable deviation of length and width of ice blocks is ±10mm.

5.5.7 The masonry of ice masonry wall should meet the following regulations:

    1 The upper and lower layer of ice masonry should be combined with staggered joints and laid internally and externally. The overlapped length should be 1/2 of ice masonry and should not be less than 120mm;

    2 The masonry height of every ice layer should be identical; a water flooding line should be sawn on the surface. The horizontal and vertical seam should not exceed 2mm, and the surface should be smooth and flat;

    3 The ice masonries (wall) of monomer ice building of the same height should be constructed in synchrony. If not, a vertical dog tooth end should be left. The dog tooth shaped wall should not be higher than 1.5m.

5.5.8 For the large volume ice landscape which adopts hollow masonry, structural methods should be adopted between the masonry, while for the non-bearing internal part, crushed ice can be adopted to fill in.

5.5.9 When it is allowed to fill the large volume of ice landscape buildings with crushed ice, the ice should be compacted, the particle size distribution of which should be reasonable; the maximum particle size should not exceed 300mm and stratification padding of the crushed ice is needed. The thickness of every layer should not exceed 1.5m. When freezing it with water, ensure it is entirely frozen. The water should not overflow the external surface of the ice masonry. The crushed ice texture should not be observed through the main facade of the ice building.

5.5.10 The ice blocks for arches should be made in a wedged mold according to the design. The deviation of upper and bottom length should be within 2mm. The width of the vertical seams should be not exceed 1mm, filled with water and must be frozen.

5.5.11 Using circular arch and the height of the wedge shaped ice arch lintel should not be less than 1/1 of the width of the hole. When the height of the ice arch is greater than 550mm, it should be divided into two layers for the masonry. The rise value should be based on the standard of Table 4.4.17-2. When the length of the ice arch hole is greater than that of the bottom of the wedge shaped ice arch lintel, the ice arch should be laid layer by layer in a staggered manner, and the staggered joint length should be 1/2 of that of the bottom of the wedge shaped ice arch lintel.

5.5.12 Holes for lamps in the ice masonry should be reserved according to the design requirements. The distance between the hole and the external surface should follow the regulation of the 4th Paragraph in section 4.6.8. And the crushed ice in the hole should be cleaned. For taller ice architecture, a concealed hole and vertical access well, in which reinforcing steel ladders should be set, should be reserved for maintenance workers.

5.5.13 All masonry surfaces of the colored ice should be smoothed. The ice seams of the colored masonry, between the colored and non-colored, should be filled with a mixture of water and colored ice powder.

5.5.14 After the construction of the outside of the ice landscape, net surface treatment should be made from top to the bottom.

5.5.15 Scaffolds and vertical transportation devices for construction must be installing individually and shall now touch with the ice masonry to avoid disturbing it.

5.6 Construction of Steel Structure in Ice Masonry

5.6.1 Seams between vertical steel reinforcements and ice blocks should be filled and frozen with ice powder mixed with water which is laid layer by layer, with vertical steel bars and horizontal hoop reinforcement should be installed in the horizontal ice trough and frozen with water but should not be higher than the ice surface or installed in the seams.

5.6.2 The seams between steel lintel, steel frame and ice blocks should be filled with water or crushed ice mixed with water.

5.6.3 The embedded parts and ice masonry should be infused and frozen with water, and there shall be no seams.

5.7 Construction of Watered Icescape

5.7.1 Framework of watered icescape should be made before construction, and only then should water be sprayed. The framework can be made at one time or made during spraying

5.7.2 The icescape can be made mechanically or manually. Spray water on frameworks made with branches or other materials, thickening the ice cover to make icicle, ice galalith, iceberg and ice cave, etc.

5.7.3 The most suitable temperature for watering icescape construction is below $-20℃$ and should not be exposed to direct sunshine.

5.7.4 Tap water and clear groundwater can both be used to make a watered icescape. The flux, strength and pulverization rate should be properly controlled when spraying.

5.8 Ice Sculpture Making

5.8.1 Ice with inclusion, bubbles or flaws should not be used for making ice sculpture.

5.8.2 Ice blocks are to be frozen together into geometric entities and then be sculpted according to the design requirement.

5.8.3 Entire ice blocks can be used for small ice sculptures. It could be sculptured after the ice blocks are built into ice base, or combined into ice sculptures after being carved. But textures and joints should meet the requirement of the works.

5.8.4 When using ice blocks to form the ice base, the seams between ice blocks should not exceed 2mm, and the infused area should not be less than 80%. The ice seams should be firmly jointed; the surface should be smooth.

5.8.5 For large ice sculptures, three-dimensional sample s can be made first, and then lines can be magnified and drawn directly on the ice base.

5.8.6 Different techniques of expression such as full relief, anaglyph, openwork carving and intaglio are all applicable to ice sculpture.

5.8.7 The ice sculpture should demonstrate the ice characteristics of transparency, refractive index, stiffness, friability and easy-weathering. It should be sculpted with clear texture and proper strength, focusing on the hollow out technique and the expressive effect as a whole.

5.8.8 According to the theme, the ice sculpture could adopt techniques of figurative or abstract. Figurative technique should represent the performance of fine, deep and vividness. Abstract technique should adopt geometric shape, physical characteristics to perform the theme and its physique characteristic.

5.9 Ice Lantern Making

5.9.1 Ice lanterns can be made into hanging-type, floor-type or other types according to different functional requirements with delicate and exquisite size. Sufficient ventilation ports should be set on the ice masonry.

5.9.2 Making ice lanterns should follow the steps as follows:

    1 Make mould according to the design requirements;

    2 Pour clear water or colored water into the mold; freeze it and the thickness of the ice billet should be between 20mm and 40mm;

    3 Empty the unfrozen water in the ice billet through digging a hole in it;

    4 Draw or sculpt on the ice crust;

    5 Set up lamps and lanterns inside the ice masonry;

6 Install auxiliary components.

5.9.3 Ice flowers can be made in the following ways:

1 Pour clear water into the mold or container and freeze it in low temperature into ice billet which is hollow inside. The forms of portrayal, carving, inlaid landscapes, fish boats, flowers, trees, old-fashioned lamps, historic buildings and characters, etc. can be adopted to reveal the theme of the icescape by drawing or sculpting inside or outside the ice blocks;

2 Pour clear water into the mold or container; put models of fish, insects, plants, flowers, small animals, which shall form into the icescapes after being frozen;

3 Pour clear water into the mold or container; mix colored solutions with different density, solubility and infusibility during the process of freezing to make special icescapes.

5.9.4 External lighting can be used for the ice flower; spotlights or other colored lights can also be used as light sources.

5.9.5 Exhibition platforms, whose height should not less than 1.0m and made of ice or other materials, should be set up in the lower part of the ice flower.

5.10 Snowscape Building Construction

5.10.1 Natural snow can be used for snow landscape building construction. Man-made snow could be used in the areas with less snow. The water ratio of man-made snow for large snow landscape building should be increased, and that for small snow landscape building should be reduced appropriately.

5.10.2 A mold of a snow billet should be firmly set up and installed layer by layer referring to the filling speed. The filling snow should be clean, and larger snow blocks and impurities are not allowed. The snow billet should be suppressed evenly and densely, and the density should follow the regulation in Table 3.2.1.

5.10.3 Snowscapes can be built by carving and shaping. Edges and corners should be smooth and the height difference of adjacent surfaces should not be less than 100mm.

5.10.4 The ornamentation of other materials on the snowscape should be firmly inlaid. Factors such as load-bearing and weathering should be taken into consideration. Large decorations should be reinforced or have an independent ground base.

5.10.5 After the completion of a small or medium-sized snowscape, surface treatment should be done to form a protecting coat.

5.10.6 Recreational facilities made of snow should be structurally strong, safe, and easily maintained.

5.11 Snow Sculpture Making

5.11.1 The snow sculpture should be designed according to the design requirements of snow base molds. The snow masonry should be compacted and the compaction should meet the design requirements.

5.11.2 The snow masonry base dimensions of art sculpture competition should be based on the game time, weather conditions, the theme of the requirements. The display effect should be reasonably determined already. In addition to the special requirements, the dimensions of length, width and height should be 3000mm×3000mm×5000mm.

5.11.3 The facade of snow sculptures should avoid direct sunlight and whole backlight. The orientations should choose illumination angle with better side light to highlight the stereoscopic sensation of snow sculptures.

5.11.4 The making process should be completed up to down and step by step, and should not make repeated adjustments.

5.11.5 The base of snow sculpture works should be used, and the base should be solid and serve as a foil to the theme landscape.

5.11.6 The performance of snow sculptures should be exaggerated, highlighting style and emphasizing outline. The sculpture form should be rough, clear with precise longitudinal and transverse lines and edges.

5.11.7 Should fully consider the influences of maintenance, conservation and weathering of the snow sculptures.

5.11.8 The illumination for nighttime viewing snow sculptures should use cold light source which placed in reasonable positions and should not influence the ornamental effect of the works.

5.11.9 The lighting fixtures should be selected according to the design requirements, rational allocating the main lights, side lights, background lights, contour lights, choose the colors meet the theme, and contorting frequencies according to the effect.

5.11.10 The regular maintenance should be conducted based on the weathering degree, maintaining a good ornamental effect.

6 Construction of Power Distribution and Illumination

6.1 Construction of Power Distribution Cable

6.1.1 Aluminum alloy cable which can work normally at or below $-25°C$ should be used and meet the insulation requirements.

6.1.2 Selection of low-voltage cables and its cross-section should meet the following regulations:

  1 Four-core cable should be used when the type of grounding is TN-C and the guard wire shares the same conductor with the neutral wire.

  2 Five-core cable should be used when the grounding type is TN-S and the guard wire and neutral wire are mutually independent.

  3 Four-core cable should be used when the grounding type is TT.

  4 When the neutral point of the electric power source below 1kV grounds directly, the cross-section of neutral conductors of quadruple three-phase cable should meet the requirement for maximum unbalanced electric current to work continuously. The effect of the harmonic current on circuit should be taken into consideration and comply with the following principles:

  1) The cross-section of neutral conductors should be no less than that of phase circuits when gas discharge lamps are the main load on the circuit;

  2) The cross-section of neutral conductors should be no less than 1/2 of that of phase circuits for other loaded circuits.

  5 When single-core cable is used as the grounding wire (PE), the cross-section area of neutral and protective conductors should match the regulation in Table 6.1.2. Cross section of neutral grounding wire should meet the following regulations:

  1) Area of copper core should be not less than $10mm^2$;

  2) Area of aluminum core should be not less than $16mm^2$.

  6 Area of cross section of protective grounding wires should meet the requirements of protecting the reliable action of appliances in circuits. It should accord with the regulations in Table 6.1.2.

Table 6.1.2 The permissible minimum area of the protective conductor qualifiedfor heat stability (mm^2)

Cross section of phase circuits (S)	The permissible minimum area of the cross section of protective conductors
S≤16	S
16<S≤35	16
S>35	S/2

  7 Since circuits of alternating current consist of many parallel cables, conductors made of the same material and with the same cross section should be used.

6.1.3 Cables should be delivered to the worksite with quality certificate, product safety certificate, product inspection report or other valid documents.

6.1.4 Appearance inspection and insulation test of cables should be made before the cables are sent to the worksite. The following regulations should be met:

1 The protective layer should not be impaired;

   2 The insulated layer should not be damaged. No flattening, distortion and loose winding of cable will be allowed. Obvious trademarks and factory marks should be labeled on the external coat of cold proof cable;

   3 Insulation test should be made; an on- the -spot test report should be completed.

6.1.5 Cable handling should meet the following regulations:

   1 Stevedoring of coil cable should be adopted when unloading. Coil cable cannot be kept flat or thrown directly off the vehicle;

   2 Non-coiled cable should be coiled to its minimum bending radius and transported after being lashed down at four points. The cable must not be dragged on the ground. Centre yarn of the cable must be sealed up with lead or insulated. It must not be affected with damp.

6.1.6 Before installation, the cable should be kept at least 24 hours in the environment with a temperature of 10℃ or higher and be set up in order.

6.1.7 Laying cable should accord with the following regulations:

   1 Cables should be inspected before laying to check if there is any damage on its surface.

   2 Cables should not be intersected when laid; they should be laid in an orderly manner and fixed. A signboard should be placed above where the cable is buried. Installation of signboards should be determined by the following principles:

   1) Signboards should be set at the beginning, ending, turning and connected joints of the cable;

   2) The line number should be marked on the signboard. Cables connected in parallel should be numbered in sequence. Handwriting on the signboard should be clear and not be easy to erase. When there's is no label incorporated in the design, types, standard, staring and destination locations should be marked.

   3 When the cable is laid, the ends of the cable should have reserve length. The direct-buried cable should keep 1.5% to 2% of the total length as redundancy, and should be laid in undee form.

   4 Protective tubes and covers should be used when cables are buried in the ground or go through the icescape. Metal steel tubes should be used when the cable might be prone to mechanical damage. The protective tube extending out of the ice architectures should be longer than 250mm.

   5 Prophylactic power cables with armour should be used in circuits of the electric power substation and box-type electric power substation, as well as all functional substations in subzones'. Steel tubes can also be added to power cables without armour and buried directly.

   6 When the cable cannot be laid beneath the scenic areas, squares or roads, the cable should be protected with galvanized steel tubes and should be covered with cracked ice which is frozen by water and not protrude above the ground.

6.2 Illumination Construction

6.2.1 Lighting lanterns should be installed according to the design requirement. The installation inside the icescape should be synchronized with the construction of the ice architecture. Electricity detection should be done accordingly, insulated measures should be adopted, and no leakage of electricity is allowed.

6.2.2 Wires beneath the ground base of the icescape should be protected with tubes. Cryophylactic insulated degree of 0.45/0.75kV should be adopted when wiring. Copper-core rubber-sheathed wire or neoprene rubber-sheathed wire should be used.

6.2.3 Thermovent should be reserved when lights are installed inside the icescape.

6.2.4 Lamps inside the icescape should be easy to install, maintain and remove.

6.2.5 The lighting inside the icescape should be integrated. Connections between two lamps should be made with module sockets or flexible connections. Joints at the power lead should be kept damp proof and encapsulated.

6.2.6 Cryophylactic electronic ballast with thermo vents shall be water and damp proof.

6.2.7 Inductive electronic ballasts should be installed intensively inside the icescape, measures of heat insulation and galvanic isolation should be taken at the bottom of it.

6.2.8 Compact energy-saving fluorescent lamps or LED lamps should be used as the point light in public areas.

6.2.9 Inside the ice masonry, LED lamps should be used with good ventilation and heat dissipation space.

6.2.10 If the ice architecture is higher than 15m or bigger than 500m^3, and an inspection channel is provided inside, a service port should be provided in the bottom or on the top of it.

6.2.11 Integrated lanterns should be installed on the stand when using project lamps or flood lights. Lamps on the stand should be able to rotate freely and make it easy to adjust the projecting angle.

6.2.12 When moldable LED lamp is used to illuminate the outline of the icescape, the distance of fixing points should be no more than 1.5m.

6.2.13 If the reactive power of gas discharge lights is too large, the diffused reactive power make-up should be done in the distribution boxes in the scenic area.

6.2.14 Illumination control in the scenic area should be controlled in situ or by using an integrated control in the duty room or electric power substation.

6.2.15 On duty and functional lighting should be provided after the scenic area is closed.

6.2.16 Power distribution wiring should conform to the following regulations:

　1 The protective grounding electric conductor (PE) should be linked with the grounding main line and a series connection is not allowed. Metal framework, the component parts of lamps and lanterns, as well as metal tubes should be grounding with marks.

　2 The color of the wire insulating layer which uses a polyphase power supply of the same landscape should be identical. The protective electric conductor (the line of PE) should choose double color wire of green and yellow. Light blue is suitable for zero line. For the color of phase conductors, A is yellow, B is green and C is red. The double color wire of green and yellow should not be chosen as load lines. The circuit mark of lighting and distribution boxes (or switchboards) should also be consistent. Load load names should be marked inside the distribution box (or switchboard) or at the bottom of the circuit breakers.

　3 Floor-type or bracket-installed lamps in crowded pedestrian paths and other places, protective measures should be taken to avoid accidental electric shock.

6.2.17 The installment of distribution boxes (or switchboards) should meet the following regulations:

1 Distribute the wires in order, and avoid twisting joints. Leading wires should connect firmly without stressing the wires. The sectional area of underlying wires on both sides of gasket bolts should be identical. No more than 2 wires are allowed to be linked to branch-wires. Components such as locking gaskets should be well prepared.

2 Switches should be flexible and reliable. Specific leakage action current with protector of residual current should be no more than 30mA, the time less than 0.1s.

3 Cylinder manifold of zero line (N) and neutral conductor (the line of PE) should be set up inside the distribution box (or switchboard). Zero line and protective conductor should be distributed through the cylinder manifold.

4 Lightning protection and grounding electrical equipments of machines that can be connected to the PE line are strictly prohibited of installing switches or fuses. The repeat lightning protection and grounding electrical equipments on machine can use the same grounding. However, grounding resistance should meet the repeated grounding resistance requirements. The working power supply connection is strictly prohibited and none circuit breaker.

5 Power and lighting circuits connected to the same load line.

6.2.18 All kinds of measuring instruments, metering devices and related electric protection (or devices) on the electrical equipment for installation, adjustment and inspection, should be quality-assured after testing. They should be in service within the period of validity.

7 Acceptance Check

7.1 General Requirements

7.1.1 The quality control of the snow landscape building construction shall meet the following requirements:

1 The main materials, semi-finished products, finished products, architectural components, instrument and equipment used in the project shall be tested. Where related to safety, energy saving, environmental protection and main function of important materials, the products should be in accordance with the provisions of the various professional engineering codes for construction and acceptance standard and design documents to conduct re inspection, and shall be subject to supervision engineer's inspection and acceptance.

2 The construction procedures should be according to the construction technical standard for quality control, each construction process is completed, the construction unit after the self-inspection in accordance with the provisions, in order to carry out the next process of construction. The related working procedures shall be carried out for inspection and recordings.

3 For the important working procedures which the supervision units ask for inspection, it should be approved by the supervision engineer in order to carry out the next process of construction.

7.1.2 When the professional acceptance of the project in the acceptance of the project do not make the corresponding provisions, the construction unit shall be organized by the supervision unit, design, construction and other relevant units to develop special acceptance requirements. Involving safety, energy conservation, environmental protection and other items of the special inspection requirements shall be borne by related personnel of the construction unit.

7.1.3 The quality control of the snow landscape building construction shall be conducted acceptance according to the following requirements:

1 Process quality acceptance shall be carried out on the basis of the qualification of the construction unit;

2 Personnel participating in the project construction quality acceptance of the parties should have the appropriate qualifications;

3 The quality of inspection lot shall be according to the main control project and general project acceptance;

4 In the construction of safety, energy conservation, and environmental protection, and the test blocks, test pieces and materials of the main use of the functions should be conducted witness and inspection in accordance with the regulations when marching into the site or during the construction;

5 Concealed works shall be carried out after the construction unit conducts inspection and acceptance as well as the acceptance documents is formed, and continue construction only after the acceptance;

6 For the structure safety, energy conservation, environmental protection and use function in the important individual works, the sampling inspections shall be carried out in accordance with the regulations before the acceptance;

  7 Appearance quality acceptance should be carried out on-site inspections by the project personnel for joint confirmation.

7.1.4 Partial and sub-projects of the snow landscape building construction should quote from Appendix D.

7.1.5 The sampling inspection of main control project quality shall all be qualified. The sampling inspection of general project shall be qualified. When using the counting sampling method, qualified rate shall be in accordance with the provisions of the code and requirements of acceptability criterion of the related professions, and there should be no serious defects.

7.2 Acceptance Check of Dominant Items of Ice Masonry

7.2.1 The intensity of ice blocks should meet the design requirements.

  Test method: Check the test report on the intension of ice blocks.

7.2.2 Pure natural water or tap water should be used for freezing water.

  Test method: Observation inspection and check the record of acceptance.

7.2.3 The structure contracture of ice blocks or staged treatment should meet the demands of the design.

  Test method: Observation inspection and check the record of acceptance.

7.2.4 The installation of expansion joints of ice masonry should meet design requirements. If not, the regulation in Parag raph 5 of Article 4.4.15 should be met.

  Test method: Observation inspection and check the record of acceptance.

7.2.5 The installation of lintels should meet design requirements. The regulations of Article 4.4.17 should be consistent with if the requirement is not stated.

  Test method: check the record of acceptance.

7.2.6 The frozen area of water crack infusion should be no less than 80%.

  Test method: Check the record of acceptance.

7.2.7 The quality of outside ice blocks should accord with the regulation of Article 5.5.1 and Article 5.5.6.

  Test method: Observation inspection and check the record of acceptance.

7.2.8 The thickness of ice wall should meet design requirements.

  Test quantity: Every inspection lot should be drawn out 10% for test; every wall space should have at least 2 tested points.

  Test method: Check by measuring with a ruler.

7.2.9 The remainder of the oblique stubble should conform to the regulation in Paragraph 3 of Article 5.5.7.

  Test method: Check the record of acceptance.

7.2.10 The width of ice seam should be not more than 2mm.

  Test method: Observation inspection and check the record of acceptance.

7.2.11 Crushed ice padding should conform to the regulation in Article 5.5.9.

  Test method: Observation inspection and check the record of acceptance.

7.2.12 The construction of curved beams should conform to the regulation in Article 5.5.10 and

Article 5.5.11.

Test method: Observation inspection and check the record of acceptance.

7.2.13 With the steel construction inside the ice masonry, the length of vertical steel bar overlap should not be less than 60d and 1200mm, the length of steel bar anchor should be not be less than 80d and 1500mm.

Test method: Check the record of acceptance.

7.2.14 The method of laying ice blocks should meet the requirements of Article 5.5.7.

Test method: Observation inspection and check the record of acceptance.

7.2.15 The length of the supportive profiled bar lintel should meet the demands of design. If no requirements are stated, it should not be less than 300mm.

Test method: check the record of acceptance.

7.2.16 The seam between reinforcing steel, profiled bar and ice should conform to the regulations in Section 5.6.

Test method: Check the record of acceptance.

7.2.17 The location of horizontal reinforcing steel should be set up according to design requirements.

Test method: Check the record of acceptance.

7.3 Acceptance Check of General Items of Ice Masonry

7.3.1 The external dimension deviation, check methods and sampling number of ice masonry construction should be accordance with the regulations in Table 7.3.1.

Table 7.3.1 The permissible external size deviation of ice masonry construction

Serial number	Item	Permissible Deviation (mm)	Test Methods	Sampling Quantity
1	Floor Height	±15	check by level gauge and ruler	not less than 4
2	Total Height	±30		
3	Surface Evenness	5	check by running ruler of 2m and wedged feeler gauge	check all the natural walls, every of which should be no less than 2
4	The Height and Width of Opening for Doors and Windows	±5	check by ruler	every test should draw out 50% and no less than 5
5	Upper and Lower Window Deviation	20	Follow the sublayer of window, check by altometer or suspension wire	every test should draw out 50% and no less than 5
6	Flatness of Horizontal Seam	7	draw a line of 10m and check by ruler	check all the external wall space, every of which should be no less than 2
7	Irregular Vertical Seam	20	check by suspension wire and ruler, follow the first peel of every layer	check all the external wall space, every of which should be no less than 2

Table 7.3.1(continued)

Serial number	Item		Permissible Deviation (mm)	Test Methods	Sampling Quantuty
8	Manhole Steps		High outside and low inside, no less than 10	draw a line and check by ruler	every test should draw out 30%, pick 3 points and should be no less than 5
9	Sideboard		±10		
10	Perpendicularity (m)	H≤15	±20	check by altometer, suspension wire and ruler	external wall, exposed corners of columns should not be less than 4, 1 for every 20m and should not be less than 4
		H>15	H/750 and ≤50		
11	Profile (axes), Length (L), Width B (m)	L(B)≤30	±20	check by altometer, suspension wire, ruler or other measuring instruments	all the external walls and internal load bearing walls
		L(B)>30	±30		

7.4 Acceptance Check of Dominant Items of Snow Masonry

7.4.1 The intensity of snow masonry should conform to the design requirements.

Test method: Check the test report of snow masonry intensity.

7.4.2 The thickness of a snow masonry wall should meet the design requirements. If no requirements are stated, the thickness of wall whose height is no more than 6m should not be less than 800mm, while the thickness of wall whose height is more than 6m but less than 10m should not be less than 1000mm.

Test quantity: Every inspection lot should be drawn out 10% for test; every wall space should have at least 2 tested points.

Test method: Check by measuring with a ruler.

7.4.3 The size of the snow section should match design requirements.

Test quantity: Every inspection lot should be drawn out 10% for test; every wall space should have at least 2 tested points.

Test method: Check measuring with a ruler.

7.4.4 The profiled bar lintel of a jack-arch opening should be match design requirements. If no requirements are stated, conformance should be in accordance with the regulations in Table 4.5.15-1.

Test method: Observation inspection and check the record of acceptance.

7.4.5 The staggered joints length of upper masonry of a profiled bar lintel should be 1/2 that of snow masonry.

Test method: Observation inspection and check the record of acceptance.

7.4.6 The length of the supporting profiled bar lintel should not be less than 400mm.

Test quantity: Every inspection lot should be drawn out 10% for test; every wall space should have at least 2 tested points.

Test method: Check by measuring with a ruler.

7.4.7 The dome-shape snow should be constructed according to design requirements. If no requirements are stated, conformance should be in accordance with the regulations in Table 4.5.15-2.

Test method: Observation inspection and check the record of acceptance.

7.4.8 The profiled bar cantilever beam should match design requirements. If no requirements are stated, conformance should be in accordance with the regulations in Article 4.5.16.

Test method: Check the record of acceptance.

7.4.9 The snow-padding quality and density should meet the design requirements. If no requirements are stated, conformance should be in accordance with the regulations in Article 5.10.2.

Test method: Check the record of acceptance.

7.4.10 The tessara construction of snowscapes should match the regulations in Article 5.10.4.

Test method: Check the record of acceptance.

7.4.11 The construction of recreational snow facilities matches the regulations of Articles 4.3.9 and 5.10.6.

Test method: Check the record of acceptance.

7.5 Acceptance Check of General Items of Snow Masonry

7.5.1 The permissible external size deviation, check methods and sampling number of snow masonry construction match the regulations in Table 7.5.1.

Table 7.5.1 The permissible external demension deviation of snow masonry construction

Serial number	Item		Permissible Deviation (mm)	Test Methods	Sampling Quantity
1	Floor Height		±15	check by level gauge and ruler	no less than 4
2	Total Height		±30		
3	Surface Evenness		5	check by running rule of 2m and wedged feeler gauge	check all the natural walls, every of which should be no less than 2
4	The Height and Width of Opening for Doors and Windows		±5	check by ruler	every test should draw out 50% and no less than 5
5	Upper and Lower Window Deviation		20	follow the sublayer of window, check by altometer or suspension wire	every test should draw out 50% and no less than 5
6	Sideboard		±10	draw a line and check by ruler	every test should draw out 30%, pick 3 points and should be no less than 5
7	Perpendicularity (m)	$H \leq 15$	±20	Check by altometer, suspension wire and ruler	external wall, exposed corners of columns should be less than 4, 1 for every 20m and should not be less than 4
		$H > 15$	$H/750$ and ≤ 50		
8	Profile (axes), Length (L), Width B(m)	$L(B) \leq 30$	±20	check by altometer, suspension wire, ruler or other measuring instruments	all the external walls and internal load bearing walls
		$L(B) > 30$	±30		

7.6 Acceptance Check of Power Distribution and Illumination

7.6.1 Acceptance Check of Power Distribution and Illumination should be accordance with the following requirements:

1 The acceptance shall be conducted by the construction unit in conjunction with the

supervision, design, construction (including sub-units), complete sets of equipment manufacturers, etc., in the construction unit on the basis of self-inspection;

  2 Power cable construction, lighting engineering construction, lightning protection and grounding should meet the requirements of this regulation, and should be completed in accordance with Appendix C;

  3 Power distribution and illumination works in the outdoor electric, variable power distribution room, main-supply, electrical power, standby and uninterrupted power supply sub-branch of engineering acceptance of sub projects should be combined with ice and snow area specific situations and the relevant professional acceptance criteria for acceptance,. The projects related to the sub projects should be implemented in accordance with the provisions of existing national standard, *Unified Standard for Constructional Quality Acceptance of Building Engineering*, GB 50300 and fill out the acceptance records according to Appendix C;

  4 Acceptance of the sub projects of power distribution and illumination should be fully accepted and approved.

7.6.2 Power distribution and illumination equipment, materials, finished products and semi-finished products in the ice and snow landscape buildings should provide quality certification information when entering into the site. For the new electrical equipment, equipments and materials should provide installation, use, maintenance and test requirements and other technical information when entering into the sit.

7.6.3 Tests of power distribution and illumination projects shall meet the following requirements:

  1 Leakage protection devices of power and lighting should be carried out of simulated action test, and make the test records.

  2 The uninterrupted preliminary operating time of energizing with full loads of large architecture lighting system within the ice and snow landscape should be no less than 24h, while that within ice landscapes should be no less than 12h and trouble free.

  3 Lamps of preliminary operate with full loads should be open. The operating situation should be recorded every 2 hours.

  4 Lamps, circuit breakers, starters, controllers, strobe and light controllers should be tested for low temperature resistance before operation. Repeated startup should not be less than 10 times and the energizing uninterrupted preliminary operating time should be more than 24 hours. Each time in the starting test of gas discharge lamps, the interval between start and stop should be not less than 15 minutes; repeated setup should be not less than 5 times. Phenomena like overheating, electricity leakage, scintillation, decay of power, exceeding starting time or abnormal starting should be avoided.

  5 With the normal operation of voltage drop, the permissible deviation of terminal voltage of power consumption equipment like lighting and electromotor (if shown by the percentage of rated voltage) is $\pm 5\%$. Monitoring surveys should be recorded.

8 Maintenance Management

8.1 Monitoring

8.1.1 Temperature monitoring of ice and snow landscapes in use should be done and the following regulations should be conformed to:

1 At least 1 representative ice and snow landscape should be selected as a monitoring target for every sector in the scenic area;

2 An icescape building higher than 12m or a snow landscape building higher than 9m should be used as the monitoring target;

3 Main structural parts or parts exposed to the sun should be selected as monitored targets. The distribution of the monitoring points should reflect the actual situation of temperature variation;

4 8:00, 14:00, 20:00 are appropriate monitoring periods. When the temperature of the monitoring point is close to the design temperature, monitoring times should be added and properly analyze the data and the deformation so as to take relevant measures.

8.1.2 The sedimentation and deformation monitoring of the ice and snow landscapes should be taken.

8.1.3 The ice and snow landscape building masonry should be maintained and removed according to the monitoring results of temperature and deformation. Corresponding maintenance measures should be taken when cracks, looseness and weathering appear locally in the ice and snow landscape building, which may affect the viewing effect by temperature, daylight and wind.

8.2 Maintaining

8.2.1 Professional and technical staff should be organized to make specific inspections of ice and snow landscapes in use. The following regulations should be conformed to:

1 Special examination should include the structural safety state of ice and snow masonry and the safe operating state of power equipment;

2 The structural safety state of ice and snow masonry should focus on deformation monitoring at an early period and on temperature monitoring in the later period. During the examination, one should monitor the temperature and deformation in the masonry's main structural parts in the monitored target;

3 The safety examination of power equipment should focus on the operating states and records of all kind of apparatus;

4 The inspection should include the operating state of impression quality, anti-skid facilities, safeguarding measures, power distribution and illumination circuits, lamps and lanterns of the distribution box or switchboard;

5 Special examination should be done once a day, while the inspection tour should be made once both before and after exhibition each day. Examinations should be increased if abnormal environmental temperature variation appear;

6 After every examination, maintenance schemes should be made on the basis of the related data and this Standard.

8.2.2 The following situations in operation need immediate maintenance:

1 The surface is polluted by snow and/or dust, etc.;

2 The ice masonry melts due to lamps inside so that holes are produced;

3 Viewing effect is influenced because of the appearance of honeycomb and pitted surfaces in the snowscape;

4 Severe weathering and partial deformation through melting. The appearance of cracks on the ice surface and absorption, wind erosion caused by adhesive joint ice seams, together with partial looseness and collapse;

5 The appearance of seams between ice or snow masonry and structural components;

6 Foundation deformation;

7 Other situations such as partial damage that affect impression quality;

8 The ice and snow landscape architecture which needs maintaining at any time.

8.2.3 Protection measures, anti-skid facilities and warning signs of recreational facilities should be maintained, strengthened or changed as needed.

8.2.4 After the completion of watering the icescape, maintenance should be made every 5 days. Water spraying should be supplemented in time to keep landscapes in good condition.

8.2.5 When the ice and snow landscape buildings reach the design temperature value for 5 consecutive days, it should be taken the measures to prohibit personnel into the upper and internal spaces, going out of service and etc.

8.3 Dismantling

8.3.1 The ice and snow landscape buildings should be dismantled when one of the following circumstances occurs:

1 If the daily maximum temperature is above 0℃ successively for 5 days;

2 If obvious displacement and slope appear which may lead to danger;

3 If the surface or parts of the landscapes are melted and the visual effect loses.

8.3.2 When the ice and snow landscape buildings are dismantled, the reusable equipment and materials should be recycled in time.

8.3.3 When removing ice and snow landscape buildings, environmental protection measures should be taken, and the environment of scenic spots should not be polluted.

8.3.4 Ice and snow landscape demolition should be based on the actual situation to take measures such as mechanical, artificial, blasting, natural melting and etc.

Appendix A Influence Coefficients of Bearing Capacity of Ice Masonry

Table A Influence coefficients of bearing capacity of ice masonry φ

Ratio of height to thickness β	Relative eccentricity $\frac{e}{h}$						
	0.00	0.05	0.10	0.15	0.20	0.25	0.30
3	1.00	0.89	0.78	0.70	0.61	0.58	0.55
4	1.00	0.88	0.76	0.68	0.60	0.57	0.54
5	1.00	0.87	0.73	0.66	0.59	0.56	0.52
6	1.00	0.86	0.71	0.65	0.58	0.55	0.51
7	1.00	0.85	0.69	0.63	0.57	0.53	0.49
8	1.00	0.84	0.68	0.62	0.56	0.52	0.47
9	1.00	0.83	0.66	0.60	0.54	0.50	0.45
10	1.00	0.82	0.65	0.59	0.53	0.49	0.44

Notes: 1 e——eccentricity of axial force;

2 h——side length of regular section which is parallel to axial force.

Appendix B   Influence Coefficients of Bearing Capacity of Snow Masonry

Table B   Influence coefficients of bearing capacity of snow masonry φ

Ratio of height to thickness β	Relative eccentricity $\frac{e}{h}$						
	0.00	0.05	0.10	0.15	0.20	0.25	0.30
2	1.00	0.91	0.82	0.71	0.60	0.53	0.45
3	1.00	0.89	0.79	0.70	0.60	0.53	0.45
4	1.00	0.88	0.76	0.66	0.55	0.50	0.44
5	1.00	0.87	0.73	0.62	0.51	0.46	0.40
6	1.00	0.85	0.70	0.59	0.47	0.42	0.37
7	1.00	0.84	0.67	0.56	0.43	0.38	0.34
8	1.00	0.83	0.64	0.53	0.39	0.34	0.31

Notes: 1 e——eccentricity of axial force;

2   h——side length of regular section which is parallel to axial force.

Appendix C Records of Engineering Quality Acceptance

C. 0. 1 The record of inspection lot is completed by a specialized quality inspector of the construction project, and the supervision engineer should arrange for the specialized quality inspector to check and record according to Table C. 0. 1.

Table C. 0. 1 The record of Quality Acceptance for Inspection _____ No. ____

Unit (sub-unit) works name		Branch (sub-branch) works name		Individual works name	
Construction unit		Project manager		Capacity of inspection lots	
Subcontracting unit		Project manager of subcontracting unit		Position of inspection lot	
Construction basis				Acceptance basis	
Acceptance items		Design requirements and standard	Minimum / actual number of samples	Inspection records	Inspection results
Major items	1				
	2				
	3				
	4				
	5				
	6				
	7				
General items	1				
	2				
	3				
	4				
	5				
Inspection results of construction unit				Professional foreman: Specialized quality inspector of project: Date:	
Inspection results of supervisor				Professional supervising engineer: Date:	

Notes: Acceptance records shall be filled out by the construction unit, and the acceptance conclusion shall be filled out by the supervision unit. Comprehensive acceptance of the conclusion should be filled in by the construction unit and agreed by the parties participating in the acceptance. Giving opinions and making comments on whether the quality of the project is in accordance with the design documents and the relevant standard and the overall quality level.

C. 0. 2 The quality of individual works should be checked by the specialized technical leader of the project organized by the supervision engineer (specialized technical manager of the project of the construction unit) and recorded according to the Table C.0.2.

Table C.0.2 The Quality Acceptance Record of Individual Works _____   No. ____

Unit (sub-unit) works name		Branch (sub-branch) works name				
Engineering quantity of individual works		Quantity of inspection lots				
Construction unit		Project manager		Technical manager of the project		
Subcontract unit		Project of subcontract unit		Subcontract content		
No.	Name of inspection lot	Capacity of inspection lot	Parts / sections	Inspection results of construction unit		Inspection results of supervisor
Description:						
Inspection results of construction unit			Specialized technical leader of project: Date:			
Inspection results of supervisor			Professional supervising engineer: Date:			

C.0.3 The quality of the branch (sub-branch) project should be checked by project manager of construction organization and related managers of prospecting and design organization organized by Chief supervision engineer (Professional technical leader of the construction unit project) and filled out in Table C.0.3.

Table C.0.3 The Record of Engineering Acceptance for Branch Works of _____ No. ____

Unit (sub-unit) works name		Engineering quantity of sub-branch works		Engineering quantity of individual works	
Construction unit		Project manager		Technical (quality) leader	
Subcontract unit		Leader of subcontract unit		Subcontract content	
No.	Name of branch works	Name of individual works	Quantity of inspection lot	Inspection results of construction unit	Inspection results of supervisor
Quality control materials					
Report of safety and functional inspection (test)					
Acceptance of Impression quality					
Conclusions of comprehensive inspection					
Construction unit Project manager: Date:		Surveying unit Project manager: Date:	Design unit Project manager: Date:	Supervisor Chief supervision engineer: Date:	

Notes: 1 Construction, surveying, design units and chief supervision engineers shall participate in and sign the acceptance of foundation and basic division of engineering;

2 The acceptance of the main structure shall be attended and signed by the construction and design unit project leaders and the chief supervision engineers.

C.0.4 Acceptance check should be recorded according to Table C.0.4-1, which is the collective table and should be used together with Table C.0.4-2 is the checking record of quality control materials. Table C.0.4-3 is the acceptance of safety and functional testing materials together with the selective inspecting record of main functions. Table C.0.4-4 is the checking record of impression quality.

Table C.0.4-1 The Acceptance Record of Completion of Unit Project Quality

Project name		Structural type		Number of layers/ area of construction	
Construction unit		Technical principal		The starting date	
Project manager		Technical principal of the project		The ending date	

No.	Project	Record of acceptance	Acceptance Conclusion
1	Acceptance of branch works	Altogether there are __ branches After the inspection, the branches which accord with standard and design requirements are __	
2	Quality control & material check	Altogether there are __ items __ items conform to requirements after inspection	
3	Safety and main function check & random inspection	Altogether __ items are checked __ items meet the requirements after inspection	
		__ items are of random inspection __ items conform to requirements after inspection	
		__ items of which are qualified after reworking	
4	Acceptance of view quality	__ items are of random inspection __ items are qualified with "excellent" and "good" __ items of which are qualified after reworting	
Comprehensive inspecting Conclusion			

Participating acceptance units	Building unit	Supervisor	Construction unit	Design unit	Surveying unit
	(seal) Project manager: Date:	(seal) Chief supervision engineer: Date:	(seal) Project manager: Date:	(seal) Project manager: Date:	(seal) Project manager: Date:

Notes: Unit acceptance of the project, the signature should be authorized by the corresponding unit of legal representative authorized by the unit.

Table C.0.4-2 The Checking Record of Quality Control Material

Project name			Construction unit				
No.	Project	Material name	Copies	Construction unit		Supervisor	
				Inspecting suggestions	Inspector	Inspecting suggestions	Inspector
1	Construction & structure	Drawing inspection, change of design, discussion record.					
2		Raw material factory certificate of acceptance and inspection (test) report					
3		Checking acceptance of concealed project					
4		Construction Record					
5		System testing and commissioning record					
6		System technical, operating and maintenance manual					
7		System management and operating personnel training records					
8		System test report					
9		Quality acceptance record of branch and individual works					
10		Records of new technology demonstration, putting on file and construction					
1	Power distribution and illumination	Drawing inspection, change of design, discussion record.					
2		Raw material factory certificate of acceptance and inspection (test) report					
3		Equipment testing record					
4		Grounding and insulation resistance test record					
5		Checking acceptance of concealed project					
6		Construction Record					
7		Quality acceptance record of branch and individual works					
8		Records of new technology demonstration, putting on file and construction					

Conclusion:

Project manager of construction unit:　　　　　　　　　　　　Chief supervision engineer:

Date:　　　　　　　　　　　　　　　　　　　　　　　　　　Date:

Table C. 0. 4-3 The Safety and Function Inspection of Survey Data and Random Inspection Record of Main Function

Project Name			Construction Unit				
No.	Project	The inspection items of safety and function		Copies	Checking Suggestions	Inspection results	Inspector
1	Construction & structure	Test report of ground bearing capacity					
2		Test report of bearing capacity of pile foundation					
3		Concrete strength test report					
4		Body structure size and location sampling records					
5		Building verticality and elevation, full height measurement records					
6		Record of building settlement observation					
7		Activity and entertainment project trial record					
1	Building electricity	Building lighting test running record					
2		Load strength test record of fixture and suspension device					
3		Insulation resistance test record					
4		Residual current action protector test record					
5		Emergency power supply time record of emergency power supply device					
6		Grounding resistance test record					
7		Pedestrian street, such as crowded places lamps anti shock measures to check records					
8		Action current and time test record of leakage protection device					
9		Sensitivity test record of electrical protection measurement instrument					

Conclusion:

Project manager of construction unit　　　　　　　　　　　　　　Chief supervision engineer:
Date:　　　　　　　　　　　　　　　　　　　　　　　　　　　　Date:

Notes: Spot checks by the inspection group of the project consultation.

Table C.0.4-4 Record of View Impression Quality Acceptance

Project name			Construction unit		
No.	Project		Quality state of random inspection		Estimation of quality
1	Construction & structure	The main structure appearance	Total ___points, ___points good, ___points average, ___points bad		
2		External walls	Total ___points, ___points good, ___points average, ___points bad		
3		Deformation seams	Total ___points, ___points good, ___points average, ___points bad		
4		Roofs	Total ___points, ___points good, ___points average, ___points bad		
5		Internal walls	Total ___points, ___points good, ___points average, ___points bad		
6		Ceiling	Total ___points, ___points good, ___points average, ___points bad		
7		Floor	Total ___points, ___points good, ___points average, ___points bad		
8		Stairs, treads, railings	Total ___points, ___points good, ___points average, ___points bad		
9		Doors & windows	Total ___points, ___points good, ___points average, ___points bad		
10		Steps, ramps	Total ___points, ___points good, ___points average, ___points bad		
1	Building electricity	Distribution box, plate, plate, junction box	Total ___points, ___points good, ___points average, ___points bad		
2		Equipment, and instruments, switches, sockets	Total ___points, ___points good, ___points average, ___points bad		
3		Lightning protection, grounding, fire protection	Total ___points, ___points good, ___points average, ___points bad		
4		Leakage protection device for distributionbox (switchboard)	Total ___points, ___points good, ___points average, ___points bad		
5		Distribution box (switchboard) N line and PE line configuration,	Total ___points, ___points good, ___points average, ___points bad		
6		Lighting quality, illumination level and effect	Total ___points, ___points good, ___points average, ___points bad		
Overall evaluation of impression quality					
Conclusion: Project manager of construction unit: Date:			Chief supervision engineer: Date:		

Notes: 1 Projects that quality evaluation is bad should be reworked;

2 Original records of impression quality site inspection should be kept as the attachment of the table.

Appendix D Division Works of Ice and Snow Landscape Buildings

Table D Division Works of Ice and Snow Landscape Buildings

No.	Branch works		Individual works
1	Foundations		Water frozen foundation, gravel foundation, assembled masonry ice foundation, wood pile foundation, steel foundation, dynamic compaction foundation, other
2	Main structure of ice masonry structure landscape construction		Ice masonry structure, reinforcement assembled ice masonry structure, steel structure, crashed ice infilling, ice arch, cantilever
3	Main structure of snow structure landscape construction		Templates, snow base, snow infilling, tessera
4	Building electricity	Outdoor electricity	Transformer; box type substation installation; whole suite of distribution cabinets; control cabinet (screen, table) and power; lighting distribution box (panel) and control cabinet installation; ladder frame, bracket, tray and slot box installation; conduit laying; cable laying; tube thread and laying within slot box; cable head fabrication, wire connection and routing insulation test; ordinary lamps installation; dedicated lamps installation; building lighting trial with power; grounding device installation
5		Substation & distribution room	Transformer; box type substation installation; whole suite of distribution cabinets; control cabinet (screen, table) and power; lighting distribution box (panel) installation; busway installation; ladder frame, bracket, tray and slot box installation; cable laying; cable head fabrication, wire connection and routing insulation test; grounding device installation; laying of grounding main line
6		Main line of power supply	Electrical equipment test and trial; busway installation; ladder frame, bracket, tray and slot box installation; conduit laying; cable laying; tube thread and laying within slot box; cable head fabrication, wire connection and routing insulation test; laying of grounding main line
7		Electrical power	Whole suite of distribution cabinets; control cabinet (screen, table) and power distribution box (panel) installation; electric motor, electric heater and power executor Inspection wiring; electric equipment test and trial; ladder frame, bracket, tray and slot box installation; conduit laying; cable laying; tube thread and laying within slot box; cable head fabrication, wire connection and routing insulation test
8		Electrical lighting	Whole suite of distribution cabinets; control cabinet (screen, table) and power distribution box (panel) installation; ladder frame, bracket, tray and slot box installation; tube thread and laying within slot box; plastic protective cased line straight laying and cabling; steel cable wiring; cable head fabrication, wire connection and routing insulation test; ordinary lamps installation; dedicated lamps installation; switch, socket and fan installation; building lighting trial with power
9		Standby and UPS	Whole suite of distribution cabinets; control cabinet (screen, table) and power; lighting distribution box (panel) installation; diesel generator unit installation; UPS device and emergency power supply device installation; busway installation; conduit laying; cable laying; tube thread and laying within slot box; cable head fabrication, wire connection and routing insulation test; grounding device installation
10		Lightning proof and grounding	Grounding device installation; lightning proof down lead and lighting arrester installation; potential connection for buildings and so on; surge protector installation

Explanation of Wording in This Standard

1 For the convenience of application of this regulation in terms of strictness, the words of different level are explained as follows:

    1) Words expressing extremely rigorous and inevitable:

        Positive: must   Negative: must not;

    2) Words expressing rigorous and should be done under normal circumstances:

        Positive: shall   Negative: shall not;

    3) Words expressing an alternative can be primarily done if the conditions permit:

        Positive: should   Negative: should not or cannot;

    4) Words expressing an alternative and can be done under given conditions: may/can/could.

2 Provisions should be specified in accordance with other relevant standard should be drafted as follows: "Shall meet the requirements of……" or "Shall comply with……".

List of the Quoted Standards

1 *Code for Design of Masonry Structures* GB 50003
2 *Load Code for the Design of Building Structures* GB 50009
3 *Standard for Lighting Design of Buildings* GB 50034
4 *Code for Design of Low Voltage Electrical installations* GB 50054
5 *Unified Standard for Constructional Quality Acceptance of Building Engineering* GB 50300
6 *Limited Values of Energy Efficiency and Evaluating Values of Energy Conservation of Ballasts for Tubular Fluorescent Lamps* GB 17896

NATIONAL STANDARD
OF THE PEOPLE'S REPUBLIC OF CHINA

Technical Standard for Ice and Snow Landscape Buildings

GB 51202 - 2016

Explanations of Provisions

Modification Explanation

Technical Standard for Ice and Snow Landscape Buildings (GB 51202 - 2016), the announcement of No. 1333 is ratified and issued by Ministry of Housing and Urban-Rural Development on October 25, 2016.

This standard is revised on the basis of *Technical Specification for Ice and Snow Landscape Buildings* (JGJ 247 - 2011). The Previous edition of the chief editorial units is Harbin Municipal Survey and Design Association. The Previous edition of the joint editorial units are Harbin Academy of Civil Engineering and Architecture, Harbin Urban and Rural Construction Committee, Harbin Architectural Designiy Institute, Ice Art Expert Committee of Heilongjiang Province, Harbin Institute of Technology, Harbin Modern Group Co., Ltd., Harbin Ark Urban Planning and Design Co., Ltd. The main drafters are Hao Gang, Wang Lisheng, Wang Dongtao, Shen Baoyin, Cao Shengxuan, Peng Junqing, Ma Xinwei, Li Jingshi, Tao Chunhui, Zhu Xiufang, Mao Chengjiu, Jiang Hongtao, Liu Ruiqiang, Liu Baizhe, Cao Lei, Sun Ying, Wang Tongjun, Wu Gang, Hao Jia, Pin Yanlin, Zhao Yiwu, Gao Guang'an, Ma Honglei, Han Zhaoxiang, Sun Guimin, Li Fu, Wang Yulei, Guo Xiangyu, Qu Huaining, Wu Fangxiao, Dong Jun and Shen Kai. The main contents of this revision are as follows: 1. Increasing the designs of Ice and Snow's structures and Masonry. The basic design calculation formula and the checking arithmetic are added. 2. Modify the foundation of ice masonry, large-body buildings of ice landscapes, round arch openings, structural components and other related construction requirements. 3. The words are standardized. 4. Deleting the contents that do not apply to display requirements.

In the process of revising the standards, after conducting extensive investigations and researches, the drafting committee not only summarized our country construction of ice and snow landscape buildings experience in practice, but also referred to some domestic and foreign technical standards. Through the materials and structure experiments of ice and snow, we get the sign and construction significant technical parameters of ice and snow landscape buildings.

In order to help the relevant staffs from various areas and organizations such as designing and construction enterprises, scientific research institutions, schools to understand correctly and implement the articles within the Standards, the code team arranges the articles in the order of chapters, sections and articles, depicts the purposes and basis of these articles as well as the cautions and warnings in using and also expounds thoroughly the reasons for the compulsory articles. However, this provision does not possess the same validity as the main body of the procedure, and only for users to understand and grasp the rules and regulations of it.

Contents

1 General Provisions ·· 64
2 Terms and Symbols ·· 65
  2.1 Terms ·· 65
3 Ice and Snow Material Calculation Indicators ··· 66
  3.1 Ice Materials ·· 66
  3.2 Snow Materials ··· 68
4 Design for Ice and Snow Landscape Buildings ·· 71
  4.1 General Requirements ··· 71
  4.2 Scenic Area Planning and Design ·· 71
  4.3 Architectural Design ··· 71
  4.4 Structural Design of Ice Masonry ·· 73
  4.5 Structural Design of Snow Construction ··· 78
  4.6 Illumination Design of Snow and Ice Landscapes ··· 80
  4.7 Intelligentization Design ·· 82
5 Construction of Ice and Snow Landscape Buildings ·· 83
  5.1 General Requirements ··· 83
  5.3 Ice-collecting and Snow-making ·· 83
  5.4 Foundation Construction of Ice Building ··· 84
  5.5 Construction of Ice Masonry ·· 84
  5.6 Construction of Steel Structure in Ice Masonry ·· 84
  5.10 Snowscape Building Construction ··· 84
  5.11 Snow Sculpture Making ·· 85
6 Construction of Power Distribution and Illumination ·· 86
  6.1 Construction of Power Distribution Cable ··· 86
  6.2 Illumination Construction ··· 86
7 Acceptance Check ·· 88
  7.1 General Requirements ··· 88
  7.6 Acceptance Check of Power Distribution and Illumination ··································· 90
8 Maintenance and Management ·· 91
  8.1 Monitoring ·· 91
  8.2 Maintaining ··· 91
  8.3 Dismantling ··· 91

1 General Provisions

1.0.1 The ice and snow landscape building is a giant leap for the arts of ice lanterns and snow sculptures. The ice and snow arts exhibition has evolved from a small-scale entertainment ornament in civil festivals to the arts of ice lanterns; further into the ice and snow landscape building, which is a comparatively long period of nearly 50 years in the North area of China. Through the chiseling and tidying by the artisans of Harbin, this exhibition has changed from a simple festival ornament into a unique art of exhibition and become an important special program to facilitate the development of the local economy and culture for many cities at home and abroad. "Ice lanterns" has also developed from the simple entertainment lanterns carried by countrymen to enjoy themselves into a comprehensive and spectacular ice and snow landscape building, with an intrinsic transformation of design, construction and function. According to the latest relevant information, there is no construction standard for materials of ice and snow in China or other parts of the world. Basing on years of practical experience and observation and numerous of tests and practical applications, as well as years of practice tests and system summaries, we revised on the basis of the original standard for reference of relative personnel.

The revision carries out the necessary improvements and adjustments of ice and snow landscape building design; supplements and refines the construction; further standardizing the work of acceptance; So as to ensure the improvement of the level of ice and snow landscape building designs, improve the construction safety and construction quality of ice and snow landscape building, and to play a role in promoting and improving the development of ice and snow art and snow culture.

1.0.2 This standard is mainly aimed at the outdoor large-scale ice and snow landscape buildings and the construction of the amusement park under the natural environment in the cold area. The design and construction of ice and snow landscape architecture in the indoor artificial refrigeration environment can also refer to this standard. Large ice and snow landscape building and its amusement park are usually constructed in the cold outside areas; therefore there are certain requirements for climate and snow and ice materials, and also certain limitations of the areas. When using this standard, it requires the combination the actual conditions of the local area.

2 Terms and Symbols

The terms and signs in the Standard are formed on the basis of the practice of design, construction and building of the ice and snow landscape buildings in cold areas in our country, the gradually cultivated customs and social recognition about ice and snow tourism and culture, and the referable information at home and abroad.

2.1 Terms

2.1.12 The low temperature in this item refers the conditions of ambient temperature which is below $-10^\circ C$.

2.1.15 The height of the ice and snow landscapes in the Standard excludes the height of the non-ice-and-snow materials in the upper and lower part of the ice and snow landscape building.

3 Ice and Snow Material Calculation Indicators

3.1 Ice Materials

3.1.1 The strength limit value of ice

1 The test curve of the limit value of the compressive strength shown in Figure 1.

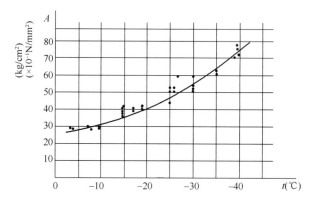

Figure 1 The test curve of the limit value of the compressive strength

The empirical formula of the compressive strength of ice is:
$$A = 26.1 + 0.24t(1 + 0.1t) \tag{1}$$
Where: A——limit value of the compressive strength of ice in different temperatures;

t——temperature of ice, taking absolute values, $5 < t < 40$.

2 The test curve of the limit value of the shearing strength shown in Figure 2.

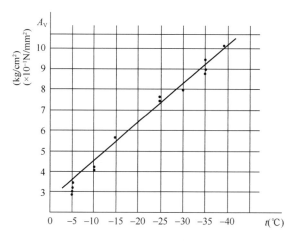

Figure 2 The test curve of the limit value of the shearing strength

The empirical formula of the shearing strength of ice is:
$$A_J = 2.6 + 0.19t \tag{2}$$
Where: A_J——limit values of the shearing strength of ice;

t——temperature of ice, taking absolute values, $5 < t < 40$.

3 The test curve of the limit value of the tensile strength shown in Figure 3.

The empirical formula of the tensile strength of ice is:

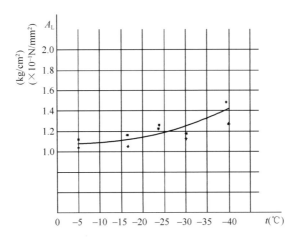

Figure 3 The test curve of the limit value of the tensile strength

$$A_L = 1.08 + 0.002t(0.13t - 1) \tag{3}$$

Where: A_L——limit values of the tensile strength of ice;

t——temperature of ice, taking absolute values, $5 < t < 40$.

3.1.2 The normal values of the compressive, tensile and shearing strengths of ice masonry

The formula of the normal value of the compressive strength of ice masonry:

$$f_k = f_m(1 - 1.645\delta) \tag{4}$$

Where: f_k——the normal value of the compressive strength of ice masonry (MPa);

δ——Coefficient of variation, 0.25;

f_m——the average value of the compressive strength of ice masonry (MPa).

$f_m = 0.52f_1$, when it is $-5°C$, $f_1 = 2.79$, f_1 is the average value of the compressive strength of ice. The data are from the test materials, namely Standard Table 3.1.1, $f_m = 1.451 \text{MPa}$.

Therefore, $f_k = 0.854$.

The formula of the normal value of the tensile strength of ice masonry:

$$f_{tk} = f_{tm}(1 - 1.645\delta) \tag{5}$$

Where: f_{tk}—— the normal value of the tensile strength of ice masonry (MPa);

δ——Coefficient of variation, 0.31;

f_{tm}——the average value of the tensile strength of ice masonry (MPa).

$f_{tm} = 0.29\sqrt{f_t}$, when it is $-5°C$, $f_t = 0.108 \text{MPa}$, f_t is the average value of the tensile strength of ice. The data is from the test materials, namely Standard Table 3.1.1, $f_{tm} = 0.095 \text{MPa}$.

Therefore, $f_{tk} = 0.047 \text{MPa}$.

The formula of the normal value of the shearing strength of ice masonry:

$$f_{vk} = f_{vm}(1 - 1.645\delta) \tag{6}$$

Where: f_{vk}——the normal value of the shearing strength of ice masonry (MPa);

δ——Coefficient of variation, 0.29;

f_{vm}——the average value of the shearing strength of ice masonry (MPa).

$f_{vm} = 0.25\sqrt{f_v}$, when it is $-5°C$, $f_v = 0.36 \text{MPa}$, f_v is the average value of the shearing strength of ice. The data is from the test materials, namely Standard Table 3.1.1, $f_{vm} = 0.150 \text{MPa}$.

Therefore, $f_{vk} = 0.078 \text{MPa}$.

When other temperatures are graded, the corresponding normal values of strength can be calculated in the same way, namely Standard Table 3.1.2.

3.1.3 The design values of the compressive, tensile and shearing strength depend on the normal value of strength. The subentry coefficient of materials (γ_f) divided by the normal value of strength is the design value of strength. Given the controllable grade (Grade C) of the constructional quality, $\gamma_f=1.8$, while for relatively concise construction, the grade can be taken as B and then $\gamma_f=1.7$.

The evaluation of the controllable grade of the constructional quality shall be in accordance with the requirements of the current *National Standard for Constructional Quality Acceptance of Ice Masonry* GB 50203, and shall be decided by factors which include the site warranty system, working environment, material strength and the overall grade level of workers' skills.

3.1.4 The formula of the ice thermal conductivity in various areas:

$$\lambda = 2.22(1+0.0015t) \tag{7}$$

Where: λ——the ice thermal conductivity [W/(m·K)];

t——ice temperature (℃), taking absolute values, $5<t<40$.

3.2 Snow Materials

3.2.1 The density of man-made snow is taken from test data and takes the reference of that from the Xuelong Brand snow-making machine instead of that of the brands of Supercool and Boton; while the density of natural snow is from the original test data.

3.2.2 The limit values of compressive strength of snow are taken from the test data. In different temperatures, the density functions of compressive strength are:

	Man-made snow		Natural snow
−10℃	$y=0.0083x-3.864$;	−10℃	$y=0.0040x-1.2113$;
−20℃	$y=0.0090x-4.1489$;	−20℃	$y=0.0043x-1.2209$;
−30℃	$y=0.0152x-7.2184$;	−30℃	$y=0.0079x-2.4415$

In these formulae, the unit of compressive strength value is MPa, while that of the density x is kg/m³.

The indexes of physical mechanics used here are test data of compressed snow, and are not true of loose snow.

The standard strength value f_k, which the strength variability of various stress states is also considered, should be accordance with *United Standard*, taking the average value of f_m when when y is taken 0.05 as the fractile in the probability distribution function. In other words, it is figured out through the formula $f_k=f_m(1-1.645\delta)$ in which f_m is taken when the guarantee rate is 95%.

Coefficient of variation $\delta=0.28$ is taken because of the relatively high discreteness of the materials.

Given the poor construction environment, when deciding the design value of strength, the controllable grade of the constructional quality is evaluated as C, γ_f, the subentry coefficient of materials, is taken as 1.9; while for relatively concise constructions, like snow sculpture, the Grade is B, $\gamma_f=1.8$. The principles to which one is to obey when deciding the design values and the controllable grades of the constructional quality are the same as that of ice masonry.

The compressive strength values under the circumstances of −15℃ and −25℃ are figured out by linear interpolation, while the values of the loose snow are excluded from the indexes of compressive strength.

3.2.3 The limit values of flexural strength of snow are taken from test data. Please see Table 1.

Table 1　The limit values of flexural strength of natural snow (MPa)

Density (kg/m³)	Temperature		
	−10℃	−20℃	−30℃
350	0.147	0.157	0.162
390	0.223	0.246	0.263
410	0.389	0.418	0.425

Man-made snow

−10℃　　　$y=0.0069x-3.3695$
−20℃　　　$y=0.0119x-5.723$
−30℃　　　$y=0.0127x-6.0505$

In these formulas, the unit of the limit value of flexural strength y is MPa, while that of the density x is kg/m³.

The calculation of standard and design values of the flexural strength of snow masonry is almost the same as that of compressive strength value in Article 3.2.2 in this Standard, with only the coefficient of variation $\delta=0.3$, the subentry coefficient γ_f respectively 1.9 and 1.8 for materials of Grade C and Grade B.

The flexural strength values under the circumstances of −15℃ and −25℃ are calculated by linear interpolation, while the values of the loose snow are excluded from the indexes of compressive strength.

3.2.4 The test data of the limit values of split tensile strength of snow masonry, the index of snow-making machine of Xuelong brand is adopted in artificial snow masonry, as shown in Table 2 and natural snow in Table 3.

Table 2　The limit values of split tensile strength of man-made snow (MPa)

Density (kg/m³)	Temperature		
	−10℃	−20℃	−30℃
510	0.093	0.113	0.121
530	0.146	0.170	0.185
550	0.194	0.216	0.231

Table 3　The limit values of split tensile strength of natural snow (MPa)

Density (kg/m³)	Temperature		
	−10℃	−20℃	−30℃
350	0.066	0.076	0.081
390	0.102	0.111	0.118
410	0.149	0.170	0.183

The calculation of normal and design values of the split tensile strength of snow masonry is almost the same as that of the split tensile strength value in Article 3.2.2 in this Standard, only the coefficient of variation $\delta=0.3$, the subentry coefficient γ_f respectively 1.9 and 1.8 for materials of Grade C and Grade B. The split tensile strength values under the circumstances of $-15°C$ and $-25°C$ are calculated by linear interpolation, while the values of the loose snow are excluded from the indexes of the split tensile strength.

3.2.5 The test data of the limit values of the shearing strength of snow masonry. Prease see Table 4, Table 5.

Table 4　The limit values of the shearing strength of man-made snow (MPa)

Density(kg/m³)	Temperature (°C)		
	$-10°C$	$-20°C$	$-30°C$
510	0.268	0.404	0.540
530	0.362	0.515	0.659
550	0.515	0.630	0.745

Table 5　The limit values of the shearing strength of natural snow (MPa)

Density(kg/m³)	Temperature (°C)		
	$-10°C$	$-20°C$	$-30°C$
350	0.068	0.072	0.089
390	0.145	0.183	0.196
410	0.179	0.200	0.221

The calculation of the normal and design values of the shearing strength of snow masonry is almost the same as that of the compressive strength value in Article 3.2.2 in this Standard, only the coefficient of variation $\delta=0.31$, the subentry coefficients are 2.0 and 1.9 respectively for materials of Grade C and Grade B considering the test value of the shearing strength is slightly higher.

The shearing strength values under the circumstances of $-15°C$ and $-25°C$ are calculated by linear interpolation, while the values of the loose snow are excluded from the indexes of the shearing strength.

According to the laboratory test data and the experience of the designers in actual works, the calculation indexes of materials of ice materials are calculated based on continuous accumulation. During nearly 50 years of practical tests, there have been no accidents occurred because of the design values of the calculation indexes of ice materials. However, the calculation indexes of snow materials have been acquired after two years' experimentation and three years' practical examinations. If necessary, more tests or refer to the experimental results can be taken.

4 Design for Ice and Snow Landscape Buildings

4.1 General Requirements

4.1.1 Overall principles are in contrast with the characteristics of short-term construction and use of the ice and snow landscape building. Great emphasis should be placed on safety and the characteristics of the local environment.

4.1.2 The paragraph 3 of this Article provides more refined and clear requirements for the names of the designs of ice and snow masonry structure.

4.1.3 The design, equipment and materials in use, maintenance and operation of the equipment and tourist activities should be guaranteed to work efficiently and effectively in the cold. New products should be tested before being put into use.

4.2 Scenic Area Planning and Design

4.2.1~4.2.5 The five articles all stipulate specific requirements of the site, overall planning, construction design and transport planning for the zones. Their implementation can refer to the design standard of the garden zones.

Each person should occupy an area of over 10m², which can still be varied by recognizing the unique features of different locations.

4.2.6 This article contains measures for emergency treatment, the implementation of the provisions of this article in accordance with the provisions of the detailed rules.

4.2.7 This article is formulated according to the requirements of National Development and Reform Commission, the National Energy Bureau, Industry and Informatization Ministry, Ministry of Housing and Urban-Rural Development

4.3 Architectural Design

4.3.1 The design of ice and snow landscape building emphasizes an external expression of the arts. Therefore, in the premise of meeting the requirements of structural safety and functions, the inside can be designed hollow and substituted by piles of soil, sandbags, and scaffolds or filled by rough or crushed ice.

4.3.2 The primary and second-primary load-bearing units of ice walls or columns over 10m high must be calculated extremely carefully and fulfill the structural requirements, particularly for those which allow tourists to enter the inner or upper part of the building. Therefore, the structure design of ice and snow landscape building is particularly important. The purpose of listing this article into compulsory standard is to establish the first safeguard for the structure design, to process the structural calculation of main and subordinate load bearing components, so as to meet the requirements of structure and texture to prevent accidents. This article must be executed compulsorily and strictly.

4.3.3 The tread width of the ice staircase takes the maximum of the specified values. It is skid-proof to design the outside of the tread higher and the inside lower; the height of the guardrails of the ice staircase takes the maximum of the national standardized values, while the thickness relies on years of practice and the ratio of height to thickness.

4.3.4 This article puts forward the requirements of "the height of ice masonry building should not exceed 30m" because of considering the elements of structure and construction, etc.; Meanwhile, given the requirements of icy masonry building in the "high", and "huge" are easy to reduce the requirements for the foundation of it; Reduce winter observations, and the costs of construction and operating; Ignore the phenomenon of meticulous elements. The design should be considered of the practical requirements, taking appropriate measures for ice masonry building that is more than 30m, so as to ensure the structural safety. Furthermore, special measures should be taken for the vertical transport of materials to ensure safety and quality management and must be intensified to ensure the delicacy and subtleness of the cutting of ice buildings.

The basis that the height of snow masonry building should not exceed 20m is:

1 Foreign technical data: height limitation of snow masonry building in Sapporo, Japan is 15m (natural snow);

2 According to the calculation;

3 According to the practical experiences, when the volume of the snow masonry reaches a certain size, the effects of temperature variation and sunlight would significantly increase. The uneven expansion will lead to severe deformation of the snow masonry;

4 If heightening is necessary, the framework should adopt the materials of steel, wood and bamboo.

Considering the temperature variation and calculation, for the building that the height exceeds 30m, there should set expansion joints no less than 20mm per 30m node.

4.3.5 The purpose of this article is to ensure the safety of construction personnel and visitors, to avoid being hurt by falling of the unstable snow objects from the top of a large ice and snow masonry. With the change of environmental temperature, wind erosion is subjected to strong wind, shock and other external forces, the large volume ice and snow landscape building may slip to threat the safety of construction personnel and visitors. Therefore, it is required that the dimensions of ice and snow masonry Must be consistent with the provisions of the geometric sizes. The seams of the masonry must meet the provisions of the bond rates and compactness in order to increase the stability of the masonry. The preparation of this article is summed up based on the occurrence of accidents in practices in order to avoid similar accidents in service phase. The Figure 4 is outline diagram of paragraph 4 of this Article. This article is mandatory and must be strictly implemented.

4.3.6 The design for ice and snow arts should introduce novel design concepts continuously and also explore original design ideas. The advanced design elements of ice and snow arts should be adopted from across the world to enrich thoughts, broaden views and make innovative developments.

4.3.7 Sunlight, temperature, wind and pollution can all melt or erode the snow. The weathering of ice is about 0.2mm a day (in Harbin area) on average, and that of ice and snow masonry will vary with different areas, environments and climates. They may even have a greater impact on snow, so if possible, it is suggested colloid sun fluid be sprayed onto the side against the sun of snow buildings or other protection measures be taken. Considering that snow sculptures should not

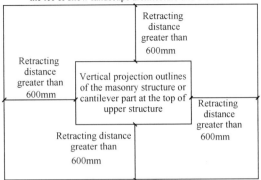

Figure 4　Vertical projection outlines of the masonry structure or cantilever part at the top of the ice & snow landscape architecture and the base

emphasize its position, however, when available, the direction of the work should be considered so as to obtain the optimum lighting effects; Increase the expressiveness of the work and reducing thawing and erosion.

4.3.8　As there are many participants in recreational activities, particularly children, safety should be the main consideration in designing such programs. Climbing events require climbing prevention measures, climbing auxiliary tools, safety protection railings at the top, evacuation platforms, and passages in case of unexpected accidents like falls, topples, skidding and medical emergencies. On the basis of years of practice and relevant design requirements, sliding events have specific technical rules on linear and curving slideway railing, the average gradients of turns and slideways, sliding tools and end design to ensure safety and reliability. The length for activities of a buffer slideway should be calculated by the gradient and the finishing point of the buffer slideway and should be installed with protection equipment. The relevant technical data in this article, covering personal safety, is in line with years of practice and must be executed compulsorily and strictly.

4.4　Structural Design of Ice Masonry

4.4.1、4.4.2、4.4.6~4.4.8　The calculation of the structural design of units of ice masonry is centered on the bearing capacity. The load effect takes the basic combination and is ensured through the corresponding structural measures.

The structure of such materials is devoid of limit values of such controlling indicators as deformation and cracking, so the calculation on the basis of their limit state in normal use will be not be that accurate and will only rely on intuitive judgment. Therefore, the calculation and the composition measures are indeed expedient to keep structures in normal use.

The relevant coefficients for Formulas 4.4.2-1 and 4.4.2-2 take values as follows:

For γ_0, the structure importance coefficient of 1.0 is used. Although the ice and snow landscape architecture will be used for a relatively short time, due to the crowded traffic and poor project stability, the structure importance coefficient of 1.0 is used;

As the *Unified Standard for Reliability Design of Engineering Structures* GB 50153-2008 recently involved the design lifetime of 5, 50 and 100 years, with the consideration of the variable load adjustment coefficient, γ_L, of structure design lifetime, this amendment list the ice masonry structure into the text as per its formula used in the load capability limit design, for the ease of visually indicating this change. Meanwhile, considering the ice and snow architecture in the natural environment of cold area can provide shorter lifetime, for the value of γ_L, 0.9 is taken uniformly;

For 1.35 S_{GK}: Taking into account the most adverse combination of the control from permanent load effect, 1.35 is used as the permanent load partial coefficient;

For 1.2 S_{GK}: Taking into account the most adverse combination of the control from variable load effect, 1.2 is used as the permanent partial coefficient;

According to *Unified Standard for Reliability Design of Engineering Structures* GB 50153-2008 Appendix A, in case there is an adverse effect on the load capability, for the value of γ_{Qi}, 1.4 is used as the variable load partial coefficient;

Under the explanations for ψ_{ci} of *Load Code for the Design of Building Structures* GB 50009-2012, any variable load in basic combination must use its combination value as representative value when working as accompanying load, and for the coefficient of the combination value, all variable load shall be 0.7, except the wind load of 0.6. Therefore, uniform combination coefficient of 0.7 is used for security.

In case of a large general landscape, it is required to identify the stratigraphic structure, geotechnical properties, hydrogeological conditions and frost depth, reaching preliminary investigation depth. For the ice and snow landscape architecture sensitive for uneven settlement, use control exploration hole.

For the ice landscape architecture with its height less than 10m, to ensure the stability of foundation, it is also required to check for the substratum under the frozen ground foundation and the basic frost heave stability. For the ice buildings built to a large scale, the venue in the early winter should have been already been covered by a layer of ice, which makes the frozen foundation soil not thick enough. Therefore, to ensure safety, before the ice and snow landscape is built on a large scale, when the frozen soil is more than 400mm thick, only 400mm is taken as the bearing layer of the frozen soil, while for the soil less than 400mm thick, the actual thickness is taken as the value. The foundation bearing capacity of frozen soil is determined by tests on site.

Static calculation scheme of open ice masonry building: When the transverse spacing(s) is greater than or equal to 20m, namely beyond the rigid analysis schemes (a temporary building will not be a heavy rigid building or roof system, so s is considered under the Purlin light building condition) in other words, in the non-rigid analysis schemes, effective composition measures should be taken to make the system become a rigid one, such as making a necessary ice transverse-wall tie or temporary wall frame which can serve as transverse-wall tie.

4.4.3 When constructing with ice block, high ambient temperature will affect the construction period and construction quality. When the temperature rises to $-5°C \sim 0°C$, the building should go out of service or be demolished. To ensure the safety of service, $-5°C$ is considered as the design temperature.

4.4.5 This article regulates that ice buildings over 10m high and with an shorter grounding side of more than 6m, basic design principles should be carried out. At the same time, for the foundation cannot meet the requirements of the design or the height of more than 10m of ice buildings, the measures should be taken against the foundation bearing capacity cannot meet the design requirements. Infusing the frozen foundation with water will supplement the requirements of checking calculation of underlying stratum and the stability of frost heave foundation, so as to assure the safety of the foundation and landscape building.

4.4.10 In this article, the coefficient of the partial pressure strength has been improved; the unfavorable construction conditions influence the construction quality and make it difficult to get an even force. Therefore, in reference to the standard end pressure of the ice masonry, the coefficient takes the round as 1.20.

4.4.11 The calculation of the load of axial tension member does not include the case that axial force is perpendicular to the adhesive surface (ice seam) between ice blocks, for example, it is to avoid such forcing pattern in design when using on-site watering for freeze as the bonding layer between ice blocks.

For the section area calculation specified in Articles 4.4.9, 4.4.11 and 4.4.12, according to related requirements of *Code for Design of Masonry Structures* GB 50003-2011, and taking into account the brittle characteristic of ice material, to ensure security, it is to calculate the stressed, axial tensile and sheared elements as per the "net" section. The same is provided for the snow structure architecture in Articles 4.5.7, 4.5.9, 4.5.10.

4.4.13 With years of practical experience, it is understood that due to poor environmental conditions and tight time limit, the construction team cannot guarantee its professional proficiency, and the case of ice seam bonding surfaces water plumpness of less than 80% happened occasionally. The masonry structure may have the possibilities of damages such as continuous seam, toothed seam and vertical seam along ice block, which have different bearing capabilities, and each has its bending tensile strength value slightly higher than the shearing strength value. Yet it is not possible to conduct bending tensile mode test, considering the need of safety, shearing strength is used alternatively.

In actual construction, both sides take the pattern of using ice for the outer wall, the watered and frozen crashed ice for filling the middle masonry structure, with the outer ice wall having its force bearing almost the same as the bending tensile element.

4.4.14 Ice buildings are for short-term sight-seeing, so they cannot be heavy buildings or roof systems that have great rigidity, so only the light purlin building or the roof system will be the horizontal bearing system of the static calculation and hence the division of rigid and non-rigid analysis schemes (including elastic and rigid-elastic). When there are heavy duty building and/or roof, it should be calculated according to actual circumstances.

For the buildings without roofs, if the ratio of sides is about or over 2, the cantilever unit will be the main consideration; if less than 2 and the three-side bearing slabs between transverse walls are high, the transverse walls should be designed into pilastered walls with ring beams or ice-column walls so that a large area of wall can be divided into lots of small slabs.

If $\frac{b}{S_0} \geq 30$ and H, the unit height of the ice wall, is the distance between two near ring beams, then H_0 can be decided according to Table 4.4.14-1, during which the rigidity of the transverse walls must reach the maximum horizontal displacement value $u_{max} \leq \frac{H}{500}$, and it should be wider than the masonry, because the materials are more shapeable, hereby, H is the total height of the transverse walls. If it is one layer, $L \geqslant H$, if more than one, $L \geqslant \dfrac{H}{2}$.

The non-rigid analysis schemes in Table 4.4.14-1 mean in particular the rigid-elastic and elastic ones, but because this project is very rare, the relevant data is not listed in detail.

The structures of ice ring beams and ice columns can refer to Article 4.4.16.

4.4.15 The couple-limb hollow ice walls often form a single limb wall when the lamps are installed in the walls and they are generally 250mm thick. In order to reinforce the rigidity and stability of structures from the early stage of construction to the later stage of use, in principle, two ice blocks should be embedded in less than 50% of height to thickness the limb limb ice wall allows along the couple-limb hollow ice walls and the steel plate web should be tied between the two ice blocks.

For the ratio of height to thickness, it refers to the joint distance between two near tied ice blocks and the thickness of thesingle limb wall.

Vertical steel bars into the drilled hole in ice columns and infusing 0℃ water with crushed ice (fragments) mixed with water and freeze them. Put the vertical steel bars into the drilled hole in ice columns and the horizontal hoops into the horizontal furrows, then pour water at 0℃ into the ice fragments to freeze them. The setting distance between the hoops should not exceed 600mm for the purpose of strengthening the binding force of the structural columns.

For the expansion joints of ice masonry, a 20mm long expansion joint every 30 meters is suggested given comprehensive consideration of structural safety, ornamental effect, and years of practice; Referencing the information at home and abroad, the expansion coefficient of ice, α, is determined as the value of 52.7×10^{-6}/K.

Major modification is made in the original standard of this article as the paragraphs 6 and 7. The reasons are as follows:

    1 The original standard of the paragraph 6 states as follows: The protective skeleton should be set up over the ice masonry whose overhanging height is greater than 0.6mm. Article 4.4.18 of the standard has stated specific provisions for its safety construction.

    2 The original standard of the paragraph 6 states as follows: The protective skeleton should be set up when round arched entrance is greater than 3m or upper part of the entrance has imposed loads. Table 4.4.17-2 of the standard has stated clear requirements for its safety construction.

    3 Practices has proved that it is not necessary for that of overhanging height is greater than 0.6mm and round arched entrance is greater than 3m when construction quality is fully accordance with the design requirements.

    4 For the round arched entrance which is greater than 3m, it will form strict constraint on ornamental effects if setting protective frame and expanded metal mesh on its external part so that the functions of transparent and translucent will not be realized.

Therefore, the modification of the protection of round arched entrance which is greater than 3m is taking protective measures such as setting expanded metal mesh or transparent separator to prevent cracked ice from falling.

Paragraph 7: Ice masonry walls are the main body of the structure of ice building, the stability of the building structure, the effect of the landscape, as well as the built-in lamp and so on. This clause makes the provisions of ice wall masonry: Internal gross amount of ice building and landscape filled with crushed ice; the thickness of exterior ice wall and blocks should not be less than 900mm or 600mm. and meet the requirements of the ratio of height to thickness of the ice wall. Ensure the integral rigidity, strength and service cycle of ice building and landscape. The limitation value of the exterior ice wall is 900mm or 600mm, which considers the condition using ice block of 600mm×300mm size. Adopting the method of flemish bond, laid layer by layer in a staggered manner.

4.4.16 For the ice and snow landscape architecture, it is allowed to refer to the temporary building anti-seismic requirements for disposal. This provision, from the anti-seismic concept, gives the anti-seismic principle. For higher ice landscape architectures, while their annual lifetime is not long, as they are to be used year after year periodically and repeatedly and with the crowded traffic, the uncertainness of earthquake occurrence and the brittleness of material lead to the cause of personal injury, so it is required to take anti-seismic structure into account, to improve the stiffness and ductility of ice structures, so as to ensure that there is no ice drop causing injury or death to visitor in case of earthquake.

As with anti-seismic structure measures, it is available to consider setting up ice-reinforced structural columns and ice-reinforced perimeter beam as well as appropriately setting up cross wall, etc., to improve structural rigidity and ductility. By increasing, it is available to increase redundancy to prevent from chain damage.

Ice perimeter beam and ice structural column are to be used as steel skeletons placed horizontally or vertically to the ice tank which has one of its sides exposed outside (a closure of three sides of ice masonry structure), and they are to be poured and refrozen to form ice perimeter beam or ice structural column with the flowing icy mixture of cracked ice and water of 0°C. It is also allowed to use other measures, such as steel frame or steel mesh perimeter beam and other anti-seismic structure.

4.4.17 Additional loads on the beam and slab: if (h_w), the height of ice wall under the beam and slab is lower than (L_n), the clear span over the beam, the load from the beam and slab is taken; while if not, the load from the beam and slab will be omitted.

The self-weight of the ice walls: if (h_w), the height of ice wall over the beam is lower than $L_n/2$, the uniformed self-weight of ice masonry will be adopted; while if not, the uniformed self-weight of half the height of walls will be adopted.

Note 3 is added into the notes of Table 4.4.17-2, providing the specific requirements of the height and rise value of the ice arch, which are important measures to maintain its overall stability. It is safer to change 1400mm to 1500mm of the rise value f_0 in the table.

4.4.18 Even if the overhanging length of the cantilever is less than 0.6m, the steel plate webs or steel bars with bars of over 0.2‰ should be installed every 1~2 layer from the above second layer and be anchored into the main structure at least $30d$.

The steel cantilever beam can choose U-bars, angle steels and double T-steel and so on.

4.4.20 When the ice buildings are over 12m high (4 stories), the rigid roof system of ice buildings should be set as the rigid transverse partition at regular certain heights(the marked ring beam) to increase the space rigidity and the whole stability of ice buildings and make the structures function coordinately. In this way every wall is furnished with horizontal supporting joints in a vertical direction to keep the walls tied.

4.5 Structural Design of Snow Construction

4.5.1, 4.5.2, 4.5.4~4.5.6 All articles should be understood and applied in reference to the relevant instructions of ice structures.

The coefficient values of formulas 4.5.2-1 and 4.5.2-2 can refer to the explanation of Article 4.4.2.

4.5.3 The snow structural units take the strength value at -10°C as the calculation index for structure design based on the following reasons. First, snow is looser than ice and becomes deformed more easily when temperature rises slightly; second, there are many factors affecting the construction quality like poor construction conditions and tight construction deadlines. When it comes to the point of dismantling at the later stage of use, the temperature will be relatively high. So to ensure safety, -10°C is the appropriate design temperature.

When calculating self-weight of snow body, the mass density should multiply the acceleration of gravity g to convert to gravitational density. For example, $550\text{kg/m}^3 \times 10\text{N/kg} = 5500\text{N/m}^3 = 5.50\text{kN/m}^3$. The snow in this article represents the snow that is pressure treated, and should take the value according to Table 3.2.1: artificial snow that the forming pressure is 0.15MPa.

4.5.7 For the snow structural units, the sections of walls and columns are all large, the walls are 800mm thick, the columns 1200mm×1200mm, usually not too high, so φ can be omitted, taken as 1. If eccentricity is a little larger, to make the snow buildings close to the axial pressed state and $\beta \leqslant [\beta]$, the sectional area can be enlarged and pilasters or frameworks can be installed.

Table 6 The coefficient of bearing capacity of snow masonry (φ)

Ratio of height to thickness, β	Relative eccentricity $\frac{e}{h}$			
	0.00	0.10	0.20	0.30
2	1.00	0.820	0.601	0.452
3	1.00	0.786	0.600	0.446
4	1.00	0.757	0.553	0.437
5	1.00	0.729	0.510	0.402
6	1.00	0.700	0.467	0.367

Note: φ is the ratio of the average value of the eccentric ultimate load to that of the axial ultimate load.

Appendix B is based on the above chart and compiled through inserting $\frac{e}{h}$ and β in a linear way. See Table 6.

4.5.8 Among any of the four cases of the calculation of the bearing capacity of the partial compressive units, namely the partial compression of the center, the middle edge of the sides of walls, the end and the corner, according to the masonry, the enhancing coefficient will not be more than 1.25. Given that snow materials are loose and will be sunken and deformed when compressed partially, the enhancing coefficient takes 1.20. In general designs the partial compressions of the end and the angle will be omitted.

4.5.9 In the calculation of the bearing capacity of the axial tensile units, the index of the axial tensile strength value is up to the split-tensile strength value.

4.5.10 In the calculation of the bearing capacity of the shear units, the index of the shearing strength is up to the according tests.

4.5.11 In the calculation of the bearing capacity of the flexural units, the index of the bent tensile strength value is up to the bent strength value, which calculated by the concentration of loads of simply supported beam.

4.5.12 The allowable ratio of the height to thickness of the walls and columns is in conformity with Table 4.5.12-2, in reference to Article 4.4.15. Only the light roof system is considered as the horizontal supporting system. Snow is weak and loose, so for walls without roofs, the cantilever structure or three-side supporting structure should be decided by the ratio of sides if the building only has three walls. When the walls are high, the design should be columns with beams or grid-like plates of ice columns.

If $\frac{b}{S_0} \geqslant 30$ and H, the unit height of the ice wall, is the distance between two near ring beams, then H_0 can be decided according to Table 4.4.14-1 (the transverse walls must have enough rigidity). The maximum horizontal displacement value $u_{max} \leqslant \frac{H}{500}$, which is less strict than masonry, because the materials are more shapeable, consequently, H is the total height of the transverse walls. If it is one layer, $L \geqslant H$, if more than one, $L \geqslant \frac{H}{2}$.

The non-rigid analysis schemes in Table 4.4.14-1 mean the rigid-elastic and elastic ones, but because this project is very rare, its relevant data is not listed in detail.

Snow materials are relatively loose and will melt and influence its stability, so itsratio of height to thickness is relatively lower than that of $[\beta]$ is allowed.

The structures of ice ring beams and ice columns can refer to Article 4.4.16.

4.5.13 The snow structures should be in conformity with the following provisions:

The structure of snow materials is relatively loose and the strength is low, and easily influenced by exposure to sunlight and wind erosion; For safety reasons, the minimum size of walls and columns are set greater, with the walls 800mm thick, the columns 1200mm×1200mm. But for the same reason, snow walls and independent columns over 10m high should be embedded with the system of bamboo, wood and steel to ensure the overall stability.

4.5.14 The conception and the arrangement measures of seismic fortification of snow can refer to Article 4.4.16.

4.5.15 The load value over the beam is taken according to Article 4.4.17.

The notes to Table 4.5.15-1 and 4.5.15-2 are only useful for the rectangular snow masonry and wedge snow blocks.

For snow arches, like ice arches, the height of each layer of wedge refers to the distance between the large and small sides of a wedge and the height of the snow arch is the total of the height of each wedge. Snow is weak and loose, subject to natural conditions, so it is necessary to check the skid stability of the arch feet, paying attention to the decrease of bearing capacity of snow due to melting, and take appropriate reinforcement measures.

Note 3 is added into the notes of Table 4.5.15-2, providing the specific requirements of the height and rise value of the ice arch, which are important measures to maintain its overall stability.

4.5.16 Because of the low shearing strength of the cantilever component of snow masonry, structural measures should be taken to secure the cantilever beam. Shape steel or other steel materials can be selected as the material of the cantilever beam.

4.5.17 Snow structure components have larger sections and better bearing capacity and stability than that of ice structures. But in higher positions, say over 9m (3 stories), because snow is easily influenced by exposure to sunlight and wind erosion, one-side melting and wind erosion will make it become an unstable eccentric compression component and easily forms an unstable load-bearing element. Therefore, the rigid roof system of ice buildings should be set as the rigid transverse partition at regular heights (the ring beam elevation) to make ice buildings stable and make the four-side walls a restrained component.

4.6 Illumination Design of Snow and Ice Landscapes

Light is the soul of the ice and snow landscape for night exhibitions. The integration of colorful and innovative light effects and the ice and snow landscape is a perfect example of the synergy between engineering technology and art displays. Light is indispensable to the design for the ice and snow landscape. Designers should have a full understanding and grasp of light, lamps, colors, power supply, and electric apparatus and lighting display; they should also make a detailed survey of new technology, new processes and new equipment of new light sources.

4.6.2 The contents and requirements of designs are as follows:

The overall design of ice and snow landscape buildings primarily includes such professional lighting design as landscape effect lighting, functional lighting, stage lighting and lighting display.

The effects of lighting mainly adopt two ways of lighting layout modes: one is within the ice, principally used in large-scale ice landscape buildings and sculptures; the other is out of the ice against ice and snow sculptures, principally used in representations of human figures, animals, plants and relief sculptures to create wonderful art effects.

Color and changes of luminance areof vital importance for the ice and snow landscape. There considerable variation in light transmittance, refraction and reflection between ice and snow, in wave length and penetration among various colors of lights, so in lighting designs and color temperature combination, it is suggested that white, red, yellow, green, blue and purple lights be used more and there is more use of contrasting or complementary colors.

In order to ensure energy conservation, environmental protection, and use frequency, light sources such as LED and fluorescent lamp should be promoted for their energy saving and low cost.

For emphasizing the lighting effects, distribute rational ambient lights on the basis of emphasis on landscape lighting. Create a good lighting scene through using functional lighting. Taking up with the position differences of lighting, and various ways such as the contrast of light and shadow, the color variation, and the combination of light spot, line, and band; Using techniques like laser, optical fiber, LED, computer program control, and laser beam 3D space modeling, etc., and ornament techniques like light sculptures, gypsophila-style lighting and red lanterns, so to achieve the perfect lighting effects.

4.6.3 According to the overall effects, the lighting design of ice and snow landscape buildings should determine reasonable color temperature to achieve the optimum effect. Favorable color rendering properties will also possess a certain energy-saving effect.

It is suggested that various light combinations are used in terms of light and lighting designs to achieve the best visual effects for the ice and snow landscape.

Glare is a tough problem for ice and snow landscape buildings and particularly impacts photography. So it is advised to arrange the lamps in a way to avoid glare. If the snow sculptures are high, high-power project-lamp, like an LED, can be used and attention given to the type of lamps and the concealing property of the light source.

4.6.4 The luminance level refers to the grades in the current national standard, *Standard for Lighting Design of Buildings* GB 50034.

The luminance range in Table 4.6.4 is decided on the basis of years of practical experience and relevant industrial standard.

In entertainment places in large ice and snow scene, lights should be used to create a happy atmosphere. Lasers can be combined with city lights, roses in the air, high-power computer project lamps to heighten the atmosphere in the scene. In addition, LED plastic lamps can be used on ice ground to form patterns and graphics, and change the light combinations.

When the scenic coversa large area, and the icescapes are greater distance apart, more functional lighting equipment outlining roads and courts should be set. Meanwhile, advertisement lamp boxes and LED lights on the ground could also be used to increase the brightness of zones.

4.6.5 When selecting light sources, rational photoelectric parameters should be determined. Lamps with excellent starting characteristics under low temperature condition should be selected.

Light sources and lamps shall meet the requirements under the low temperature conditions.

For large-scale ice and snow landscapes, due to the limitations of the site, lifting equipments may not be able to get close to them, therefore the quality of lamps must be under strict control.

The overall design of the garden lights need to rationally distribute the luminance among each zone to avoid large contrast of color and brightness. For the cast lighting (floodlighting) of ice and snow landscape, determine the luminance or brightness of various parts of the facade of the illuminated objects. Equally illumination of the whole landscape is not advised. However, phenomenon of obvious spots, dark areas or distortions could not be found in the same lighting area as well.

4.6.6 Flourescent lamps are widely used in the current large-scale ice and snow landscape buildings. During the demolition, they will be torn down together with the landscape instead of being recycled. When the lamps are broken, the glass fragments, mercury and poisonous substances will blend into ice and causing pollution to the environment. Therefore, the use of environmental protection, easily recycling, and reusable light sources is advocated.

4.6.7 Landscapes where hold major activities should enhance its load rating of power supply.

The load of three-phase voltage should be kept balanced and the difference among each phase voltage should not be excessive.

The important lighting load should be evenly divided into two special circuits, namely to the electric powers to simplify the system and reduce the automatic switchover levels.

The general lighting load is mainly single-phase when three-phase circuit-breakers are used. If one single breaks down, it will make a three-phase tripping operation and increase the power-out areas.

The relevant rules are mainly based on an unbalanced lighting load and the higher harmonic out of the non-linear circuit of gas discharge will make the neutral conductors also able to pass the current at odd times of 3, which will then reach the phase value. Therefore the relevant provisions have been made.

Ordinary circuit-breakers (including micro breakers) are properly used at above 5℃. In cold areas, select the product that can properly operate below −30℃.

Single-phase grounding protection should be adopted for personal safety protection.

For outdoor cabinets and boxes, when electric components give off heat, ice and snow snow and ice on shells will melt into cabinets and boxes. Therefore, necessary protective measures should be taken.

4.6.8 The light inside the ice landscape shall be color-harmonious and arranged skillfully and fancifully. Nowadays T5 tri-phosphor tube lights are recommended as internal lamp, as the tube is thin and the pre-provided grooves in the ice are small and convenient for construction. It is suggested LED plastic lamps or other energy-saving and environment-friendly lamps be widely used for they are free of halogen, highly efficient, have a long life after installation, strong penetration, low power consumption, and convenient maintenance.

New energy-saving, environmental protection, high efficiency, safety, low temperature resistance, low cost light source is the development direction of snow and ice landscape architecture. At present, the LED lights product specifications or standards, there has been no uniform, no sound technical specifications, products used in the process influencing factors need to be considered.

4.6.9 For the luminance requirements of the façade of the snowscape, diverse colors of halid floodlights, high power LED floodlights and animated lights can be utilized to project a stereovision of the snowscape, to give life to the lifeless snow and create a fantastic ice world.

4.7 Intelligentization Design

4.7.1~4.7.4 Promote the level of people-oriented service and scientific management through the Internet system. Other service items include: Provide web promotion, online ticketing, regional monitoring and intelligent systems for the scenic area.

5 Construction of Ice and Snow Landscape Buildings

5.1 General Requirements

5.1.1 The technical disclosure of ice and snow landscape building consists of design disclosure and joint checkup of construction drawings. The technical disclosure includes the design requirements of the ground base, major structures, non-ice supporting structures, internal or external lights, external landscape modeling and external landscape not shown in the working drawings. Emphasizing on special issues in ice-now construction, problems encountered should be solved during the technical disclosure.

5.1.2 This article means that on the basis of technical disclosure, the construction unit should make optimal selection of structure construction scheme and construction method, preparing the design of construction organization and audit according to the provisions.

5.1.3 Ice building over 30meters high and snow building over 20meters high are focus of monitoring objects. For their complicated structures, the requirements of the foundations and integral rigidity are relatively strict. To avoid the structural variations caused by climate influence, exclude the strengthening managements of design and construction, continuous observation of settlement and deformation must be conducted in the process of using. For example, when settlement crack or severely defamation of ice and snow masonry are found, protective measures such as consolidation and partial closure should be taken; Set up a safe distance, non-construction personnel shall not enter the danger zone. This article involves not only the safety during the construction period; it also involves the security after putting into use. This article is mandatory and must be strictly implemented.

5.1.4 To ensure structural safety and function, it is very necessary to check important materials and equipment before use. This article aims to address any ignorance or avoidance of regulations which may occur when the work is tough and deadlines are tight.

5.1.5 The surface of ice and snow landscape should be processed to improve visual effects and mitigate the erosion.

5.3 Ice-collecting and Snow-making

5.3.1 This article focuses on the area where ice is in greatest need and which is able to supply ice.

According to tests of materials and practices, when natural ice is less than 200mm thick, it will be too weak to endure the operational weight and will be prone to accidents. Then the ice will be too easily broken to be molded.

The raw ice to be used contains amounts of water i.e. the ice has not been formed totally. To meet the strength requirements, raw ice should be kept cold for a while for it to be completely frozen. After being cut, "rough ice" forms six standardized and smooth blocks for construction.

The dimension regulation of ice materials aims at regulating and standardizing the design of ice masonry. The geometric size of natural raw ice is based on years of practice for the convenience of site processing and reducing wasted ice.

5.4 Foundation Construction of Ice Building

5.4.3 To guarantee the structural safety and stability of ice buildings, the wall must locate on the base by virtue of the entire ice masonry, especially the large ice building and platform whose insides are filled with crushed ice, the upper side of which must be piled up from the base to the top. It is strictly prohibited from locating the ice wall and icicle on the filled crushed ice cover.

5.5 Construction of Ice Masonry

5.5.2 To ensure the landscape is clean and tidy, clean natural water or tap water should be used to water ice seams.

5.5.5 The temperature of masonry varies with the environmental temperature during construction. Temperature monitoring of the constructed masonry should be done accordingly to control its temperature. If it is higher than $-5℃$ (the design temperature value), construction should be terminated and related measures should be taken to ensure safety. This article is mandatory and must be strictly implemented.

5.5.7 The laying masonry during the construction should meet the requirements of this article. If existing conditions are that the affused cohesion water is easier to flow when the ice seam is too large, the ice seam should not exceed 2mm.

5.5.9 Large volume ice landscape building should be made by laying ice blocks during the construction. When the design requirements allow filling with crushed ice, relevant provisions of this article should be met.

5.5.12 This article regulates that the distance between the hole of lamps and lanterns and external surface of ice masonry is regulated to be not more than 350mm and not less than 150mm, considering the effect of transmittance, layout of in-built lamps and ice melting damage caused by increase of temperature, projective position of the sun and the appearance of areolation. The optimum location, the density, lighting intensity of lamp layout should meet the requirements of ice transmittance and design effects through the results of actual tests.

5.5.15 Scaffolds and vertical transportation devices for construction are strictly prohibited from installing on the ice landscape building and shall now touch with it to avoid disturbing its masonry and the complete of outer surface because of external force. During the construction, relevant measures for stabilizing scaffolds should be taken. Dual rows of steel tubes with closed hand-lap cross-ring frames should be used for scaffolds. Place the ice landscape building in the middle of the scaffolds and rigid connections shall be designed between them.

5.6 Construction of Steel Structure in Ice Masonry

5.6.1~5.6.3 Measures should be taken to ensure tight joints between reinforcing steel or steel structure and ice blocks, for example, adopting the method of infusing the joint by a mixture of crushed ice and water as well as freezing firmly. The horizontal stirrup embedded into the tank should not be higher than ice surface and is to meet the construction requirements.

5.10 Snowscape Building Construction

5.10.1 Man-made snow should be used in the areas where the snow column is limited. Its water ratio is related to the degree of snow density and intensity, which is determined after on-the-spot tests.

5.10.2 The neatness of snow should be in the appearance of the snow landscape, the specific requirements of which are regulated in this clause. As for snow-blank making, model shaping is suggested and separated layer tamping is recommended to ensure density and intensity of snow.

5.10.4 This article are the requirements for tessera of the snow landscape building.

5.11 Snow Sculpture Making

5.11.1 The emphasize on compactness is to ensure the reliable structure of the main body.

5.11.2 Dimensions of snow blocks are based on years of summed up practical experience. The main referenced comprehensive factors are the number of participants, the production time limit, the use of tools, physical strength, environmental impact and etc. Dimensions of snow blocks of other large and medium scale snow sculptures depends on concrete conditions, but the structural measures should meet the requirements of relevant provision of the standard.

5.11.3~5.11.9 Requirements from perspective of arts: It should be comprehensively considered about layout of lights, color deployment and etc.

6 Construction of Power Distribution and Illumination

6.1 Construction of Power Distribution Cable

6.1.2　The number of cable cores and the choice of sections should give priority to safety and appropriateness.

6.1.3~6.1.5　The requirements of the entrance and delivery of cables into factories aims at the quality and safety of construction.

6.1.6, 6.1.7　Cable laying should give priority to the protection of cable safety, while taking into account the economy.

The fixed power supply system: the cables are fixed all year round, supplying power in winter and summer, with supply to other appliances in summer.

Temporary power supply system: the cables are laid for winter display and dismantled after use according to the requirements.

6.2 Illumination Construction

6.2.1　In case of reworking of dismantling the ice, the installation of internal lanterns should be synchronously with the construction of the ice masonry companying with random tests for checking whether the electrical equipments can normally start or flicker. The electrical equipments internal the icescape must take measures of moisture and water proof. No leakage allowed so as avoiding the melting of ice, forming the electrical conductor.

6.2.2　The pipes and the wires can be laid simultaneously in wiring under the icescape in case ice blocks crush them. Low temperature resistant rubber wires or chloroprene rubber with copper cores can be used.

6.2.3　Heat dissipation shall be noted and done when quantities of inductive ballasts are set together.

6.2.4　Measures should be taken in design and construction to make it easy to recycle the lamp tubes and wires.

6.2.5　Connection accessories should be used in integrated lanterns to be easily installed. Meanwhile, moisture proof and sealing treatment should be paid attention to of wire connections.

6.2.10　Maintenance port for lamp replacement should be set in large volume ice landscape building. The ports can be set according to specific requirements under the premise of structure safety assurance.

6.2.13　The casting light (floodlight) is gas luminaire. When centralized employment of it is taken, compensation of capacitor in the nearby distribution box could be done caused by the influence of low power factors.

6.2.14　The integrated remote control system can be used for the illumination of the large scale landscape in the distribution rooms or duty rooms to uniformly control on and off switching. Light control, combined with time control, is preferred when the street lamps or the garden lamps are used as the duty guard lamps.

6.2.16 The protecting grounding (PE) near naked conductors of electric equipment or conduits should be firm and reliable in case of electricity leakage that might cause harm. The grounding branches and mains should never be connected in series in case when one component is dismantled, the other grounding branches or null individual lines, fail to protect from electric shock.

The colors of the outside layers of wires are set to differentiate their function and make it easier to recognize, maintain and examine. PE cannot be used as a load wire. The colors of the insulating layer of the wires should vary with different functions in the same spot. The lighting circuit in the spot should be in line with that in the power distribution box with load names to recognize, maintain and examine so as to avoid accidental electric shock due to improper operation.

With the constant improvement of the ice and snow arts, the types of external projecting lights have increased. The lamps, if installed in crowded places, are easily touched, so there should be strict anti-burn and anti-shock measures.

6.2.17 There should be less than 2 wires at each connection end to ensure closeness and that no loosening of the wires occurs through heat expansion and cold contact after being charged.

To cleanly divide PE and N wires, TN-S system is utilized and PE and N rows are respectively laid in the power distribution boxes.

There are differently rated capacities in the power distribution boxes and a few circuits exist in small capacity, only 2~3. They can be combined respectively into PE and N wires through several connection columns, but no mixed connection is allowed.

6.2.18 Whether the indication and signs of instrumentation dials are correct directly influences the judgment of the working state, expected function and safety requirements. Therefore, this article is a specific rule.

7 Acceptance Check

7.1 General Requirements

7.1.1 According to the current national standard *Unified Standard for Constructional Quality Acceptance of Building Engineering* GB 50300, the constructional quality acceptance of ice and snow landscape building is divided into units (sub-units) works, branch works, individual works and inspection lots.

First of all, units (sub-units) works in the ice and snow landscape can be divided, based on the independent construction and appreciation function in different function divisions, among which, the single works of independent appreciation function or independent spots can be regarded as sub-unit works. If it is decided by the building unit, supervisor and construction institutions before construction, then collect the data and acceptance it.

Branch works should be determined by specialty nature and architectural parts. If the branch works are heavy and intricate, the works which have the same parts or can form independent specialty systems can be divided into several sub-branch works. The branch works can be further classified into foundation and ground, body structures as well as power distribution, and lighting sub-branch works. The main body structures branch works contain sub-branch works of ice masonry, steel structure, crashed ice filling, mould, snow blocks and etc. The power distribution and illumination branch works contain sub-branch works of cable laying, internal lamps, external lamps, distribution wiring, safety protection, and power distribution box installation.

Individual works are made of one or several inspection lots. The inspection lot can be divided into construction parts, deformation cracks based on the construction quality control, and acceptance. Every three meters of ice masonry works, snow masonry works, ice masonry structure works and steel (wood) structure works inside can be regarded as one inspection lot.

Design of activities of ice and snow, barrier-free design, design of safety facilities, scenic area service management design, supporting facilities design (service provider, water supply, drainage, power supply, heating, sanitation facilities and identification) should be done according to the requirements of this regulation and related professional acceptance criteria, in which the facilities construction quality acceptance should be strictly implemented in accordance with the design documents and the requirements of relevant standard.

Concealed works of foundation and ground, steel (wood) structure inside the ice and snow masonry, equipment inside ice masonry and cable construction should inform the designing, supervising and building units to check before concealing and then forming the document of concealing acceptance.

Second, the quality acceptance of the inspection lots (individual works) of ice-snow landscape building is the key step of the engineering quality, and the important means to guarantee it. Before the acceptance, the construction unit should fill out in "Quality acceptance record of inspection lots and individual works", and the inspector and professional technical leader should sign their signatures respectively. After that, the supervision engineer should organize and conduct the acceptance in strict accordance with the procedural provisions.

The individual work acceptance will be conducted under the responsibility system of general supervision engineer (technical leader of construction unit) by the project leader, technical and quality leader of construction unit. Since subgrade and foundation, technical performance of major structure, electricity protection, lighting operation commissioning are related to the safety of the whole works, therefore the project leader of the design unit should be required to take part in quality acceptance of relevant individual works and be responsible for the acceptance results.

After the units (sub-units) works of the ice-snow landscape buildings is completed, the construction unit should firstly organize relevant personnel to conduct self-inspections based on the quality standard and design drawings, and assess the inspection results. If the works is qualified, the acceptance report should be handed over with the complete quality materials to the construction unit for recheck.

The unit works acceptance should be organized and designed by the construction unit or project leader, inspected and accepted by leader of construction unit, or project leader and technical and quality leader of construction unit, the general supervision engineer of supervisor and technical leader of operation and management unit. The participation of the operation and management unit is to repair relevant defects before putting into service and to maintain and manage during the service.

Third, construction unit works acceptance of ice and snow landscape building, which is also called quality completion acceptance, is the last acceptance before buildings are put in use and is also the most important one. Besides the qualified acceptance of the individual works constituting the unit works, all the relevant documents and files should be complete; there are also inspections should be conducted as follows:

1 The recheck of inspection materials involving safety and function use of sub-branch works involves not only the integrity inspection (no missing checking items), but also the witness sampling inspection reports supplemented into the acceptance of the individual works. Besides the completion of the relevant documents and files, inspections should also be conducted as follows:

2 Spot checks should be conducted on main function, which is a comprehensive inspection of the the ultimate quality of the ice and snow landscape building construction, equipments and lamps installation. On the basis of qualified individual and sub branch works, a thorough examination should be done in the completion inspection. The items of spot checks should be determined jointly by personnel participating in the acceptance with sampling methods of measuring and counting. The checks should be conducted as the requirements of this standard.

3 It is necessary that all relevant participants of acceptance together making an instinctive quality inspection. Such inspection is difficult to apportion to those present and can only be made by methods of observing, touching and simple measurement and be judged by each person's subjective impression. The results do not show "qualified" or "unqualified", but a comprehensive quality evaluation. For evaluations deemed "bad", the constructions will be returned for repair.

Forth, the unqualified circumstances will be identified and solved in the acceptance of inspection lots, which demonstrates the principle of simultaneous construction, inspection and correction. Due to the tight construction period of the ice and snow landscape building, all potential quality dangers should be erased in the construction of inspection lots as soon as possible.

When the construction quality does appear defective, they should be handled as follows:

1 During the acceptance of inspection lots, when the structural programs fail to satisfy the requirements or the dimensional deviation is not qualified, measures should be taken properly. If the functions and safety were influenced, they should be demolished for reworking. Common defects should be corrected by repairing or replacing the equipments. The construction unit is allowed to review the inspection lots after according measures, but only after one inspection lot is checked qualified, can another be made. Quality safety can never be ignored due to the tight construction period.

2 For the unqualified inspection lots, if the original design department checks and identifies them still qualified for safety and function use, they can be accepted.

3 For inspection lots with severe defects, when handled according to certain technology plans, they can still be used safely, which means they change the structural size of the shape, but if safety and function are maintained they can still be accepted according to the technical plans and negotiation documents.

7.1.2 For the branch works and engineering units not qualified for safety after return-repair or reinforcing, they must be dismantled and can never be put back into use.

7.6 Acceptance Check of Power Distribution and Illumination

7.6.2 It is crucial for the construction quality to make an entrance inspection of major equipment, materials, finished products and half-finished products. There should be records of the procedures and inspection results (including valid certification reports) and they should be confirmed by relevant iunits. Checks and inspections of the new electric equipments, instruments and materials should be made before their entrance to the site to ensure the later relevant procedures work smoothly.

7.6.3 Descriptions of relevant articles:

1 To avoid the malfunction of electric equipment, converting the equipment to the charged body close to the bare conductors and result in electric shock accidents, the leakage protection device should be installed to immediately cut off the electricity to avoid accidents. There should be simulation tests on the leakage protection device to ensure its sensitiveness and reliability.

2 The test time of full load charge should be regulated, which is an effective way to check whether the electricity in the scenic area can work normally in peak times.

3 All lamps should be checked one by one to ensure the intact rate.

4 Check the stability of all electric equipments.

5 All electric equipments have certain requirements on voltage variation. If the variation is beyond that allowed, the life span of the electric equipments will be shorter or will reduce the light flux.

8 Maintenance and Management

8.1 Monitoring

8.1.1 There is a direct relationship between temperature and strength of the ice and snow landscape buildings. The masonry strength will vary with temperatures. Consequently, the specific requirements of temperature measurement of ice and snow masonry are made.

8.1.2 Besides the temperature measurement of ice and snow masonry, monitoring of structural deformation in operation should be taken simultaneously. For the strength of ice and snow landscape buildings is not only bound up with temperatures, but foundation, construction, weathering and wind erosion as well. The structural deformation in operation reflects a comprehensive result of relevant elements and is very important for safety application.

8.1.3 This article puts forward relevant measures when the 3 monitoring results appear.

8.2 Maintaining

8.2.1 In accordance with experiences, areas where are above 40 degrees north latitude or at higher altitudes, the service life of the ice and snow landscape usually lasts: 1. from late December to early January for initial application period. 2. from late February to early March for the later. The time limit for the functions of the buildings mainly depends on the influences caused by temperature variations, the melting of the buildings and the judgments of the application effects. The time limit could refer to the natural conditions, environmental conditions and design requirements of the local area.

8.3 Dismantling

8.3.1 Through years of practical experiences and temperature measurements and deformation monitoring on the spot, the specific requirements of dismantling the ice and snow landscape buildings are made.

8.3.2, 8.3.3 It is required that recycle the reusable equipments and materials. Meanwhile, measures should be taken to meet requirements of environmental protection.